STATISTICAL METHODS WITH APPLICATIONS TO DEMOGRAPHY AND LIFE INSURANCE

STATISTICAL METHODS WITH APPLICATIONS TO DEMOGRAPHY AND LIFE INSURANCE

Estáte V. Khmaladze

A revised and expanded version of the Russian edition
Editor of the English translation: Dr Leigh Roberts
Translator: Dr Mzia Khmaladze

CRC Press
Taylor & Francis Group
Boca Raton London New York

CRC Press is an imprint of the
Taylor & Francis Group, an **informa** business

A CHAPMAN & HALL BOOK

CRC Press
Taylor & Francis Group
6000 Broken Sound Parkway NW, Suite 300
Boca Raton, FL 33487-2742

First issued in paperback 2019

© 2013 by Taylor & Francis Group, LLC
CRC Press is an imprint of Taylor & Francis Group, an Informa business

No claim to original U.S. Government works

ISBN-13: 978-1-4665-0573-5 (hbk)
ISBN-13: 978-0-367-38023-6 (pbk)

Visit the Taylor & Francis Web site at
http://www.taylorandfrancis.com

and the CRC Press Web site at
http://www.crcpress.com

Contents

Sections with an "*" are supplementary material.

Preface

The idea of this book first crossed my mind in the early 90s, when, following the collapse of the Soviet Union, so many good mathematicians, especially those outside probability theory and mathematical statistics, found themselves unable to earn their bread by proving theorems or even teaching mathematics. "But here it is," I thought – "such a fine field of applications as demography, and such a useful area as life insurance. Why not make the transition from pure mathematics to these fields easier?" Reading demographic books, notably, those of R. Pressat, H. U. Gerber and N. Keyfitz, as well as learning numerical demographic realities, helped to shape the mathematical program of our first seminars and lectures.

The content of the book was finally decided upon and used at University of New South Wales and now at Victoria University of Wellington, as a one semester course. It centers mostly around analysis of an individual life: although we speak about samples of many lifetimes, the range of problems is still that pertaining to an individual life. The statistical methods it describes are those based on empirical and related processes.

Although the text is mostly mathematical, the book discusses things outside mathematical theory, especially if we understand theory in a narrow and technical sense. The language is kept informal and almost colloquial. There is no organization into theorems and lemmas, yet exact mathematical statements are formulated and proved, albeit as economically and cleanly as I could. Each statement, and each object introduced, is illustrated, and its presence in the book justified, by statistical problems they help to solve.

The few initial lectures are relatively simple and will be easy to read, but some other material may require a little bit more effort to learn. Overall, this is an attempt to speak about not totally elementary methods in clear and concise form. In doing this I tried to bring the book closer to the modern day statistics.

I also tried to make the range of topics relatively diverse. In particular, Lecture 15 speaks about the analysis of the tails of distributions of life-times, while the next lecture, Lecture 16, presents a model for population dynamics with migrations. Some separate sections, for instance, those on the Stieltjes integral and the Wiener integral, are intended to help with technical points. Some others, like those on mixtures of distributions, extreme value theory and the age structure of a population, introduce other themes of interest in demography. Lectures 12, 13 and 14 form a somewhat separate group and discuss net premiums for various insurance policies. At the same time, the attempt has been made to keep the presentation short in order to use the book for a half year course.

A systematic study of statistical problems concerning portfolios of insurance policies and age-specific population dynamics models is planned for another book.

At the initial stage of my work within demography and insurance mathematics I enjoyed the support of Prof. N. Vakhania, who helped to formally establish the course in mathematical demography at Tbilisi State University, and Dr G. Mirzashvili, now Chief Actuary of Aldagi BCI, the largest insurance company in Georgia, who was a very staunch supporter of these activities. The role of the Demographic Institute of the Georgian Academy of Sciences and its director of the time, Prof. L. Chiqava, was remarkable: although the institute was traditionally very non-mathematical, as was all of Soviet demography, instead of the more usual resistance to change, we had from them full support for the new program.

It is important for me to thank those who helped in preparation of this book. For the Russian edition it was Tatiana Tolozova and then mostly Maia Kvinikadze who prepared the TeX file, and Mzia Khmaladze, who translated the whole text into English and helped to polish the TeX file. Leigh Roberts spent a very large amount of time on the text, helping, with great care and patience, far beyond an editor's duties. Ray Brownrigg prepared all the graphs and carried out many numerical experiments, which helped to refine the text. He and Gennadi

Martynov provided tables for Lecture 7, as did Eka Shinjikashvili for Lecture 15. My students, Ian Collins, and Yoann Rives read the text and made useful comments. Especially this is true for Yoann Rives, who read the text extremely carefully, as well as one could wish one's book is read by anybody: he discovered misprints, a mistake and suggested improvements. One more such reading was given to the book by my current PhD student Thuong Nguyen. Very useful discussions I had with my colleagues David Vere-Jones and John Einmahl, with John – particularly around Lecture 15. On analysis of longevity in Japan I enjoyed discussions with Masaaki Sibuya and on mixtures, with Max Finkelstein.

I am also grateful to John Kimmel of Chapman & Hall. I do not know if I would've writen the book without his attention to the project and careful encouragement.

Estáte Khmaladze

List of Figures

List of Tables

Duration of life as a random variable

Duration of life of an individual person, and that is the subject of our book, is considered in mathematical demography as a random variable.

This, of course, is not the only possible point of view on duration of human life. There is something simplistic and vulgar in such a point of view. In M. Bulgakov's *The Master and Margarita* (see English translation Bulgakov [1992]), in a dialog between Pontius Pilate and Yeshua, there is the following passage:

"What would you have me to swear by?" enquired the unbound prisoner with great urgency.

"Well, by your life," replied the Procurator. "It is high time to swear by it because you should know that it is hanging by a thread."

"You do not believe, do you, hegemon, that it is you who have strung it up?" asked the prisoner. "If you do you are mistaken."

Pilate shuddered and answered through clenched teeth:

"I can cut that thread."

"You are mistaken there too," objected the prisoner, beaming and shading himself from the sun with his hand, "You must agree, I think, that the thread can only be cut by the one who has suspended it?"

From these words of Yeshua, which we regard with great reverence, it is clear that for the one who has suspended the life by the thread, the duration of life is not a random variable at all.

But for us, humans, this is perhaps the only possible point of view. We cannot imagine how we could possibly know about biochemical reactions in any particular human body, or how many eggs and of

what quality he would consume, and subsequently how much choles-
terol would accumulate in his blood vessels, or what would pass in the
minds of drunken drivers that he may come across in his life, or what
would be the rain and wind that might push the brick from the rooftop
and drop it on his path, and where this path might lie. And if so, we
are left with no option but to perceive the duration of life as a random
variable.

What we gain as a result of this perception is an extremely flexible
and highly practical theory to describe the behavior of human populations.

And so, let a random variable T denote duration of life of a person.
In probability theory, a complete description of a random variable is
given by what is called its distribution function:

$$F(x) = P\{T < x\}, \qquad 0 \leq x < \infty.$$

Everyone would agree that for any $x < 0$ the probability $P\{T < x\} = 0$, so that probability that duration of life will be less than -3 years
or -1.5 years is equal to 0. True, that one could be interested, for
example, in durations of pregnancy, still placing 0 at the moment of
the birth, and this would create the values of $x \in [-9 \text{ months}, 0]$, but
we will not be considering this in the present lectures and once and for
all agree that the distribution function

$$F(x) = 0 \quad \text{at all} \quad x \leq 0.$$

Everyone would also agree that if $x_1 \leq x_2$, then $F(x_1) \leq F(x_2)$:
e.g. the probability of not reaching the age of 35 years is less than
the probability of not reaching the age of 47 years. Finally, as $x \to \infty$,
the distribution function $F(x) \to 1$. Indeed, at $x = 200$ years $F(x)$ is
practically 1, and more so at $x = 1000$ years.

Therefore, the distribution function $F(x)$ is a function of age x with
the following properties:

$$F(x) = 0 \quad \text{for} \quad x \leq 0,$$
$$F(x_1) \leq F(x_2) \quad \text{for} \quad x_1 \leq x_2, \qquad (1.1)$$
$$F(x) \to 1 \quad \text{as} \quad x \to \infty.$$

Any function of x with these properties can serve as a distribution function of a duration of life. In other words, duration of life is simply a non-negative random variable.

Although $F(0) = 0$, the limit of $F(x)$, when x tends to 0 from the right,

$$F(0+) = \lim_{x\downarrow 0} F(x),$$

in general, does not have to be equal to 0 and the saltus $F(0+) - F(0) = F(0+)$ is equal to the probability of still birth. We can, however, agree to consider durations of life of only live birth individuals, and the distribution function of these durations has to be continuous at 0: $F(0+) = F(0) = 0$.

It is often more convenient to talk about the complement of a distribution function to 1,

$$F^*(x) = 1 - F(x) = P\{T \geq x\},$$

which, at each x, gives us the probability that a person will survive until the age x. It is clear that

$$F^*(x) = 1 \quad \text{for } x \leq 0,$$
$$F^*(x) \downarrow, \tag{1.2}$$
$$F^*(x) \to 0 \quad \text{for } x \to \infty.$$

In survival analysis and demography F^* is usually called a survival function.

The expected value, ET, or mean, of duration of life T is defined as the integral

$$ET = \int_0^\infty x F(dx) = -\int_0^\infty x dF^*(x).$$

If this integral is finite — and why would one use distribution functions with infinite expected values to model duration of human life? — then integration by parts leads to the equality

$$\int_0^\infty x dF(x) = -x[1 - F(x)]\Big|_0^\infty + \int_0^\infty [1 - F(x)] dx$$

$$= \int_0^\infty [1 - F(x)] dx. \tag{1.3}$$

Although $x[1 - F(x)] = 0$ for $x = 0$ is always true, in (1.3) we need to be also sure that $x[1 - F(x)] \to 0$ if $x \to \infty$ as well. However, this is indeed true if ET is finite: since $\int_0^\infty x\, dF(x) < \infty$, then $\int_z^\infty x\, dF(x) \to 0$ for $z \to \infty$, and since $\int_z^\infty x\, dF(x) \geq z[1 - F(z)]$, then $z[1 - F(z)] \to 0$ when $z \to \infty$ as well. Therefore, if $ET < \infty$, then

$$ET = \int_0^\infty [1 - F(x)]\, dx = \int_0^\infty F^*(x)\, dx. \qquad (1.4)$$

This form of representing an expected value is very commonly used in the demographic literature.

The variance of T is defined as the integral

$$\mathrm{Var}(T) = \mathsf{E}(T - \mathsf{E}T)^2 = \int_0^\infty (x - \mathsf{E}T)^2\, dF(x).$$

As we know, the variance is a measure of spread of T around its expected value: the higher the value of the variance, the larger this spread, and the lower the variance, the higher the concentration of T. If we had a random variable, which is always equal to its expected value, then its variance would be 0.

Distribution function, expected value, variance – these are very general characteristics, applicable to any random variable. Now we shall introduce different characteristics that are specific to T being a duration of life, as opposed to, say, T being a random weight or random distance.

The first such characteristic is the conditional probability of a person surviving until age $x + y$ (let it be, say, event A), given that the person survived until the age x (and let that be an event B). If the probability of B is not 0 (if, that is, x is not equal to 1000 years), then the conditional probability of the event A given event B is defined as

$$\mathsf{P}(A|B) = \frac{\mathsf{P}(A \cap B)}{\mathsf{P}(B)}. \qquad (1.5)$$

However, the intersection of events $A = \{T \geq x+y\}$ and $B = \{T \geq x\}$ is, certainly, the event A itself. Therefore

$$\mathsf{P}\{T \geq x+y \mid T \geq x\} = \frac{\mathsf{P}\{T \geq x+y\}}{\mathsf{P}\{T \geq x\}} = \frac{F^*(x+y)}{F^*(x)}. \qquad (1.6)$$

The conditional probability that a person will not survive until age $x + y$, given he has survived until age x, by the same rule (1.5), equals

$$P\{T < x+y \mid T \geq x\} = \frac{P\{x \leq T < x+y\}}{P\{T \geq x\}}$$

$$= \frac{F(x+y) - F(x)}{1 - F(x)}. \qquad (1.7)$$

Probability (1.7) as a function of y has all the properties of a distribution function, listed in (1.1): for $y < 0$ it equals 0, it is non-decreasing in y and it converges to 1 as $y \to \infty$. In other words, (1.7) defines a distribution function in y. Let us use shorter notation for it:

$$F(y|x) = \frac{F(x+y) - F(x)}{1 - F(x)}. \qquad (1.8)$$

The complement of $F(y|x)$ is, obviously, the probability defined in (1.6):

$$F^*(y|x) = 1 - F(y|x) = \frac{1 - F(x+y)}{1 - F(x)}. \qquad (1.9)$$

As a function of x, $F(y|x)$ can behave in a more or less arbitrary way and is nothing like a distribution function in x. The role the variables x and y play in (1.8) and (1.9) is asymmetric.

Another important quantity, which characterizes T, is the expected value of remaining life. This is simply an expected value calculated using the conditional distribution function $F(y|x)$:

$$E[T - x \mid T \geq x] = \int_0^\infty y F(dy|x).$$

Again, using integration by parts as we did in (1.3), we obtain

$$E[T - x \mid T \geq x] = \frac{\int_0^\infty [1 - F(x+y)] \, dy}{1 - F(x)} = \frac{\int_0^\infty F^*(x+y) \, dy}{F^*(x)}. \qquad (1.10)$$

The total expected value of T, given a person lives until the age x, is, certainly, equal not to (1.4) but to the expression derived in (1.10) plus x:

$$E[T \mid T \geq x] = E[T - x \mid T > x] + x = x + \frac{\int_0^\infty F^*(x+y) \, dy}{F^*(x)}. \qquad (1.11)$$

In the case of a live birth, i.e. if $F^*(0+) = 1$, and we agreed to consider only such cases, we see that the limit of $E[T | T \geq x]$ as $x \downarrow 0$ equals (1.4).

It is worth noting that as a function of x the expected duration of life $E[T | T \geq x]$ is increasing. If, for example, expected duration of life at birth is 75 years, then for a person who lives to the age of 30 years, it is greater, and for a person who reached the age of 52 years, it is still greater. How much greater? That depends on the form of the distribution function F.

◇ **Exercise.** Show that (1.11) defines an increasing function of x. Use the fact that $F^*(x)$ is non-increasing and do not worry about differentiation under integral sign.   △

◇ **Exercise.** Derive the expression of (1.11) explicitly, when $F(x)$ is uniform distribution on $[0, \omega]$ with some "ultimate age" ω, that is, when $F(x) = x/\omega$ for $x \leq \omega$ and $F(x) = 1$ for $x > \omega$.   △

◇ **Exercise.** Repeat the last exercise, when $F(x)$ is is the so-called exponential distribution function, that is, when $F(x) = 1 - e^{-\lambda x}$ for $x \geq 0$ and where $\lambda > 0$ is a parameter of the distribution. Is the result intuitively plausible in a demographic context? What is the value of the expected duration of life for $x = 0$?   △

The fact that the expected duration of life is increasing with age was intuitively clear long ago. One example of this, popular in actuarial and demographic literature, is the example of Ulpian's tables of expected duration of remaining life (see, e.g., Haberman and Sibbett [1995], Kopf [1926]).

Roman citizens have been supplied by the state with grain, olive oil and some other products; this was called an "aliment". Ulpian needed such tables for forecasting the necessary stock of these products. Although these tables were used some 14 centuries before any serious work in probability theory had begun, the understanding that $E[T | T \geq x]$ should be increasing in x is very much present there: it was stipulated that a $20 - 24$-year-old Roman citizen would receive aliment, on average, for another 27 years, that is, it was assumed that an

Age	1–19	20–24	25–29	30–34	35–39	40
Aliment	30	27	25	22	20	19
Age	41	42	43	44	45	46
Aliment	18	17	16	15	14	13
Age	47	48	49	50–54	55–59	60+
Aliment	12	11	10	9	7	5

Table 1.1 *Ulpian table of duration of aliment, given to Roman citizens*

expected lifetime of such a citizen would be $47 - 51$ years; to Roman citizen of the age $50 - 54$ years the aliment would be given, on average, for another 9 years, that is, the expected lifetime of such a citizen would be $59 - 63$ years.

Ulpian was senator and lawyer during the reign of Emperor Alexander Severus (222–235) . He left a considerable legacy in Roman jurisdiction. According to Edward Gibbon [1776], Ulpian was an uncompromising and active statesman. In 228 he was murdered by soldiers of the Pretorian Cohort (cohort stationed in Rome) in the presence of the Emperor, a youth of 18-19 years old, to whom he appealed for protection.

By considering small y in the distribution function (1.8) we will arrive at one of the main characteristics of the duration of life. We prefer to write Δx, a "small increment" of x, instead of y and $\Delta F(x)$ for the corresponding increment of $F(x)$:

$$\Delta F(x) = F(x+\Delta x) - F(x).$$

Then we will have that

$$\frac{\Delta F(x)}{1 - F(x)} \qquad (1.12)$$

is the probability of death "soon after" or " immediately after" reaching the age of x. Suppose, our distribution function $F(x)$ has a so-called density, which we denote as $f(x)$, that is, assume that

$$F(x) = \int_0^x f(y)\,dy.$$

Actually, only distribution functions with densities will be used as distribution functions of T. Then we can speak about the limit

$$\mu(x) = \lim_{\Delta x \to 0} \frac{\Delta F(x)}{\Delta x} \cdot \frac{1}{1 - F(x)} = \frac{f(x)}{1 - F(x)}. \tag{1.13}$$

This is the most important demographic characteristic, called the force of mortality. Force of mortality is the density of the distribution function $F(y|x)$ at $y = 0$. In general reliability applications it is called a failure rate.

Previous formulae represent expected remaining life and force of mortality through the distribution function $F(x)$. Note, however, that the other way round, if we know one of these functions, we can reconstruct $F(x)$ and therefore the other function as well. It is particularly useful to note how $F(x)$ is represented by $\mu(x)$: if $\mu(x)$ is a function such that

$$\mu(x) \geq 0, \qquad \int_0^\infty \mu(x)\,dx = \infty,$$

then

$$F(x) = 1 - e^{-\int_0^x \mu(y)\,dy} \tag{1.14}$$

is a distribution function and has a density.

Thus, any non-negative function of age x, such that its integral from 0 to ∞ is infinite, can be used as a force of mortality.

◇ **Exercise.** Will the function in the right-hand side of (1.14) remain a distribution function if any one of conditions on $\mu(x)$ is not satisfied? △

◇ **Exercise.** If instead of the rate of mortality we were given the expected remaining life for each age x, could we reconstruct the distribution function F? The answer is yes: denoting, for simplicity of notations, $e(x) = \mathsf{E}[T - x \mid T \geq x]$, so that $e(0) = \mathsf{E}T$, show that

$$1 - F(x) = \frac{e(0)}{e(x)} e^{-\int_0^x (1/e(y))\,dy}.$$

To show this, consider (1.10), or

$$e(x) = \frac{\int_x^\infty [1 - F(z)]\,dz}{1 - F(x)},$$

as an equation with respect to F and solve it. △

The integral

$$\Lambda(x) = \int_0^x \mu(y)\,dy \qquad (1.15)$$

is also a frequently used characteristic in reliability theory. It is called the risk function; see, e.g., Barlow and Proshan [1975]. We will come across risk functions in Lectures 8 and 9.

Now let us agree on some terminology we may use later on.

Namely, let us agree that we may sometimes call the duration of life T the lifetime, expected value of the duration of life ET we may call expected life, while the expected value of the remaining life $E[T - x | T \geq x]$ at the age x we may call expected remaining life without direct reference to the age.

1.1* A note on Stieltjes integral

Let $g(x)$ be a continuous function of x and let $F(x)$ be a distribution function for $x \geq 0$. We want to define the integral $\int_0^\infty g(x)dF(x)$.

First consider the integral on a finite interval $\int_0^c g(x)dF(x)$. Start with partitions of this interval by n point $0 = t_{1n} < t_{2n} < \cdots < t_{nn} = c$ and assume that these partitions become more and more fine everywhere on $[0, c)$ when n increases: $\max(t_{k+1,n} - t_{kn}) \to 0$ as $n \to \infty$. Let $\Delta F(t_{kn})$ denote an increment of F on the interval $[t_{kn}, t_{k+1,n})$, $\Delta F(t_{kn}) = F(t_{k+1,n}) - F(t_{kn})$. Then we define $\int_0^{c-} g(x)dF(x)$ as the limit

$$\int_0^{c-} g(x)dF(x) = \lim_{n\to\infty} \sum_{i=0}^n g(x_{kn})\Delta F(t_{kn}), \quad x_{kn} \in [t_{kn}, t_{k+1,n}).$$

As soon as we know what the integral is for finite c, we can define the integral $\int_0^\infty g(x)dF(x)$ as its limit:

$$\int_0^\infty g(x)dF(x) = \lim_{c\to\infty} \int_0^{c-} g(x)dF(x).$$

If the distribution function F has a derivative (which is its density) at all x, then the increments $\Delta F(t_{kn})$ are all tending to 0 as $n \to \infty$, but they can be approximated by the differentials $f(t_{kn})[t_{k+1,n} - t_{kn}]$ with high accuracy:

$$\Delta F(t_{kn}) = f(t_{kn})[t_{k+1,n} - t_{kn}] + o(t_{k+1,n} - t_{kn}),$$

and as a result,

$$\sum_{i=0}^{n} g(x_{kn})\Delta F(t_{kn}) = \sum_{i=0}^{n} g(x_{kn})f(t_{kn})[t_{k+1,n} - t_{kn}] + o(1).$$

But the limit of the sum on the right side is just the Riemann integral. Therefore, if F is differentiable at all $x \in [0,c]$, then the Stieltjes integral coincides with Riemann integral:

$$\int_{0}^{c-} g(x)dF(x) = \int_{0}^{c} g(x)f(x)dx.$$

Imagine, however, that F at a point x is discontinuous: $F(x+) - F(x) > 0$. For each n this point belongs to one of the intervals $[t_{kn}, t_{k+1,n})$. Take a sequence of such intervals. Then the corresponding summands $g(x_{kn})\Delta F(t_{kn})$ will not tend necessarily to 0, but to $g(x)[F(x+) - F(x)]$. The overall limit, however, may still very well exist. Hence, Stieltjes integral is defined even if the Riemann integral is not.

◇ **Exercise.** Verify that if F were a discrete distribution function, concentrated at points $0 \le x_1 < x_2 < \dots$ then

$$\int_{0}^{\infty} g(x)dF(x) = \sum g(x_k)[F(x_k+) - F(x_k)]$$

which is exactly an expression for the expected value $Eg(T)$ for T with a discrete distribution. Although we will not use discrete distributions as a model for the distribution of durations of life, it is still useful to understand this fact.   △

Models of distribution functions $F(x)$ and force of mortality $\mu(x)$

In Lecture 1, the duration of life was introduced as a positive random variable T with some, in fact basically any, distribution function $F(x)$. As soon as we do not require anything specific from $F(x)$ we are facing an enormously rich choice from the class of all distribution functions on $[0, \infty)$. However, from practical analysis of data we know that some distribution functions are much more appropriate and relevant than others.

Nevertheless, we do not want to say that distribution functions that fit the data most accurately are necessarily the best. We believe it is much more important to have, or to develop, some qualitative assumptions concerning T, to derive an analytic form of distribution functions based on these assumptions, and then to investigate the agreement between these distribution functions, that is, between our qualitative assumptions and the real data. As an illustration, let us recall a classical example from nuclear physics: the model for radioactive disintegration.

In the process of radioactive disintegration, or α-decay, so called α-particles are spontaneously emitted from a radioactive substance. The emitted particles can be more or less accurately recorded by a counter. The time intervals between successive emissions are obviously random and could, it may seem, have any distribution function. Let us assume, however, that

(a) the emission of an α-particle from a nucleus is completely spontaneous, without any "preparation" within the nucleus;

(b) the emission of one α-particle does not influence the emission of other particles: that is, an emission of one α-particle does not "excite" other nuclei;

(c) the probability of emission of an α-particle during a short interval of time $[t, t+\Delta t]$ depends only on the length of this interval and is asymptotically proportional to Δt; that is, it has the form

$$\lambda \Delta t + o(\Delta t),$$

and the coefficient λ is constant.

The assumptions here are quite non-trivial. Why, for example, should particles not be emitted in clusters, by twos or threes? And why should the flight of one particle not excite other nuclei? And why, indeed, should there be no "preparation" for emission within a nucleus? But none of these can happen under our assumptions. In fact it follows from these assumptions that the distribution function of times between successive emissions must be

$$F(x) = 1 - e^{-\lambda x}, \qquad x \geq 0, \tag{2.1}$$

and cannot have any other form. Therefore, testing for agreement between the distribution function (2.1) and the experimental data is not so much fitting something to something but the testing of our qualitative assumptions (a)–(c); see Rutherford et al. [1930], Marsden and Barratt [1910], Marsden and Barratt [1911].

We do not know of an example so vivid and of such fundamental importance within mathematical demography. None the less, *a priori* qualitative assumptions also play an essential role in the selection of appropriate distribution functions F for T.

The choice of F is principally based on the properties of the corresponding force of mortality μ. This aspect of mathematical demography does not differ from general reliability theory and survival analysis – reliability of no matter what, of a human organism, or a mechanical tool, or an electronic gadget. As we said in Lecture 1, within survival analysis μ is called a failure rate. If $\mu(x)$ is increasing in x, that is, if the probability of an immediate failure at reaching an age x increases with this age, then we have the process of "ageing". If $\mu(x)$ is decreasing in x, then we have the process of "hardening" – see the case of the lognormal distribution (2.13) below. Distributions of this latter type are not really useful for demographic problems, but can be quite appropriate in certain reliability problems.

The case when

$$\mu(x) = \lambda \quad (= \text{const})$$

plays a specific role. As immediately follows from formula (1.14), the distribution function corresponding to this $\mu(x)$ is the exponential distribution function, which we have seen above in (2.1). Therefore, to say that failure happens spontaneously, without "preparation", is equivalent to saying that the force of mortality, or failure rate, is constant.

The exponential distribution proves to be useful when describing mortality from accidents. Quite unexpectedly, it was discovered that the duration of rule of Roman emperors is also well described by the exponential distribution (Khmaladze et al. [2007]). This example will be discussed more fully in Lecture 6 . In reliability theory a more detailed classification of distributions is used, rather than just those with increasing or decreasing failure rates. Here are the main groups in this classification (see, e.g., Barlow and Proshan [1975]):

1) distributions with increasing failure rate: $\mu(x) \uparrow$;

2) distributions with increasing failure rate in the mean: $\frac{1}{x} \int_0^x \mu(y)\, dy \uparrow$;

3) distributions with decreasing expected remaining life: $E[T - x | T > x] \downarrow$;

4) "new is better than used":

$$P\{T - x > y | T > x\} < P\{T > y\};$$

5) "new is better than used, in expectation":

$$E[T - x | T > x] < ET$$

All these groups reflect the heuristic idea of ageing and wear and tear.

The most simple and yet interesting case for "regular" demographic problems is the case of increasing $\mu(x)$ – people do in fact age. This is illustrated by Figure 2.1 which shows a statistical estimate of the force of mortality for the New Zealand population (http://www.stats.govt.nz/). (For statistical estimates see Lectures 3 and 6.)

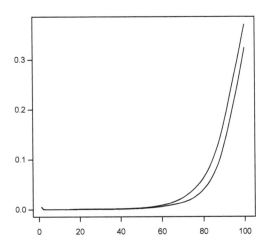

Figure 2.1 *Graph of force of mortality μ for male (continuous line) and female population of New Zealand in 2003–2005.*

Choosing $\mu(x)$ as the following power function

$$\mu(x) = k\lambda^k x^{k-1}, \qquad \lambda > 0, \ k \geq 1, \ x \geq 0, \qquad (2.2)$$

brings us to the so called-Weibull distribution (Weibull [1939]):

$$F(x) = 1 - e^{-(\lambda x)^k}, \qquad x \geq 0. \qquad (2.3)$$

Here $\lambda > 0$ is a scale parameter of the variable x. A more pessimistic model than the one described by (2.2) is the model of "exponential ageing" – the so-called Gompertz model, (see Gompertz [1825]):

$$\mu(x) = \theta c e^{cx}, \qquad \theta > 0, \ c \geq 0, \ x \geq 0. \qquad (2.4)$$

According to (2.4) it is not only the force of mortality that increases, but the rate of increase as well, and this rate is proportional to $\mu(x)$ itself:

$$\mu'(x) = c\mu(x).$$

The Gompertz distribution function corresponding to (2.4) is given by

$$F(x) = 1 - e^{-\theta(e^{cx}-1)}. \qquad (2.5)$$

In both the Weibull and Gompertz cases the force of mortality is a convex function, as in Figure 2.1: that is, the derivative $\mu'(x)$ is increasing – "ageing accelerates".

In fact it is not true that a monotonically increasing force of mortality is the most realistic case. People working in demography know that, in particular, the force of mortality is subject to periodic (yearly) variations see e.g. (Landers [1993]). We do not have, as far as we know, established analytical models with such periodically variable $\mu(x)$, which increases only in the mean (as in group 2 above). But we do have methods of non-parametric analysis/estimation of $\mu(x)$ that will be discussed in Lecture 7.

One can consider models more optimistic than (2.2). According to one such model, the force of mortality $\mu(x)$ increases, not indefinitely, but up to some finite limit k. If, in particular, μ is proportional to $F(x)$, then we will have

$$\mu(x) = kF(x). \tag{2.6}$$

This is in fact a differential equation for $F(x)$:

$$F'(x) = kF(x)[1 - F(x)]. \tag{2.7}$$

Solving this equation yields the so called logistic distribution function:

$$F(x) = \frac{1}{1 + e^{-k(x - x_0)}}. \tag{2.8}$$

The logistic distribution function can be used to describe the distribution of duration of life for, say, $x \geq x_0$. Note that x_0 is the so called median age:

$$P\{T \geq x_0\} = P\{T \leq x_0\} = F(x_0) = \frac{1}{2}.$$

At $x = 0$ the distribution function (2.8) exceeds zero, and therefore it is not a distribution function of a positive random variable. However, the numerical value $F(0) = (1 + e^{kx_0})^{-1}$ for realistic values of k and x_0 is often small, and may generally be ignored.

The so-called gamma distribution function is defined by its density

$$f(x) = \frac{\lambda^k x^{k-1}}{\Gamma(k)} e^{-\lambda x}, \qquad x \geq 0,$$

where λ is the scale parameter and

$$\Gamma(k) = \int_0^\infty x^{k-1} e^{-x} dx$$

is the gamma function of "shape parameter" k. The gamma distribution function is the integral

$$F(x) = \frac{\lambda^k}{\Gamma(k)} \int_0^x y^{k-1} e^{-\lambda y} dy, \qquad x \geq 0. \qquad (2.9)$$

For the gamma function see, e.g., Abramowitz and Stegun [1964]; for the gamma distribution function see, e.g., Johnson et al. [1994] and Johnson et al. [1995]. When $k = 1$, F becomes an exponential distribution function while for $k = n/2$ with integer n and $\lambda = 1/2$ it becomes a chi-square distribution function with n degrees of freedom. We use both of these distributions in this book.

For integer values of k the integral in (2.9) can be calculated explicitly, and $F(x)$ can be written as the sum

$$F(x) = \sum_{j=k}^\infty \frac{(\lambda x)^j}{j!} e^{-\lambda x}, \qquad k \text{ - an integer.} \qquad (2.10)$$

This, however, is not very important for us at present. It is more important to understand the behavior of the corresponding force of mortality

$$\mu(x) = \frac{x^{k-1} e^{-\lambda x}}{\int_x^\infty y^{k-1} e^{-\lambda y} dy}, \qquad (2.11)$$

especially for large values of x. It is not always possible to express $\mu(x)$ through elementary functions, unless (2.10) is true; that is, unless k is an integer. However, we can always calculate it and look at its graph.

◇   **Exercise.** Investigate to what extent the distribution function (2.9) is suitable to describe the populations with empirical force of mortality similar to the one shown on Figure 2.1.
a. The expected value of the distribution function (2.9) equals k/λ, while its standard deviation equals $\sqrt{k}/\lambda$ (that is, the variance equals k/λ^2). Find appropriate values of parameters k and

λ, choosing the expected value as equal to, say, 75 years and the standard deviation as equal to, say, 10 years.

b. Using R or Excel, or any software you want, calculate values of $\mu(x)$ for $x = 30, 40, \ldots, 90$ and draw its graph. Does it look like the graph in Figure 2.1?

c. Does this graph agree with the asymptotics $\mu(x) \to \lambda$ for $x \to \infty$, which we will derive below? $\triangle$

Let us investigate the asymptotic behavior of (2.11) for large values of x, or more formally,

for any $k > 0$, when $x \to \infty$ the force of mortality of the gamma distribution function (2.9) converges to λ.

Indeed, rewrite (2.11) as

$$\mu(x) = \left(\int_x^\infty \left(\frac{y-x}{x} + 1 \right)^{k-1} e^{-\lambda(y-x)} \, dy \right)^{-1}$$

$$= \left(\int_0^\infty \left(\frac{u}{x} + 1 \right)^{k-1} e^{-\lambda u} \, du \right)^{-1}.$$

In the last integral, the function of u, which we integrate, monotonically tends to $e^{-\lambda u}$ as $x \to \infty$ for every value of u. Therefore

$$\mu(x) \to \left(\int_0^\infty e^{-\lambda u} \, du \right)^{-1} = \lambda. \tag{2.12}$$

Note that although $\mu(x)$ behaves asymptotically as a constant for any gamma distribution, it is equal to a constant only for the exponential distribution.

$\diamond$ **Exercise.** Recall that

$$1 - F(x) = e^{-\int_0^x \mu(y) \, dy}$$

and, therefore

$$1 - F(y|x) = \frac{1 - F(x+y)}{1 - F(x)} = e^{-\int_x^{x+y} \mu(y) \, dy}.$$

Using these equalities show that convergence of (2.12) implies that

$$\frac{1 - F(y|x)}{e^{-\lambda y}} \to 1$$

for every y and for $x \to \infty$, although it does not imply that

$$\frac{1 - F(x)}{e^{-\lambda x}} \to 1.$$

Intuitively, this means that if durations of life followed gamma distributions, the mortality at high ages would become more and more spontaneous, as if following an exponential distribution. $\triangle$

A distribution that is very frequently used in many fields of research to describe positive random variables is the so called lognormal distribution. A random variable X has a lognormal distribution if $\ln X$ has a normal distribution. In other words, if $\Phi(z)$ denotes the standard normal distribution function,

$$\Phi(z) = \frac{1}{\sqrt{2\pi}} \int_{-\infty}^{z} e^{-\frac{x^2}{2}} \, dx,$$

and $\ln X$ has a normal distribution with expected value m and variance σ^2, then the lognormal distribution function is just

$$F(x) = P\{X \le x\} = P\left\{ \frac{\ln X - \mu}{\sigma} \le \frac{\ln x - \mu}{\sigma} \right\}$$

$$= \Phi\left(\frac{\ln x - \mu}{\sigma} \right). \qquad (2.13)$$

Differentiating with respect to x we see that its density is given by

$$f(x) = \varphi\left(\frac{\ln x - \mu}{\sigma} \right) \frac{1}{\sigma x},$$

where $\varphi(z) = \frac{1}{\sqrt{2\pi}} e^{-\frac{z^2}{2}}$ denotes the density of the standard normal distribution function.

The lognormal distribution function is interesting for us as an example of a distribution function with decreasing force of mortality/failure rate. Indeed, using the asymptotic formula (see, e.g., Feller [1971] or Abramowitz and Stegun [1964]),

$$1 - \Phi(z) = \frac{1}{z} \varphi(z)[1 + o(1)], \qquad z \to \infty,$$

we obtain that

for the failure rate of the lognormal distribution

$$\mu(x) = \frac{\frac{1}{\sigma x} \varphi\left(\frac{\ln x - \mu}{\sigma}\right)}{1 - \Phi\left(\frac{\ln x - \mu}{\sigma}\right)}$$

the following asymptotic representation is true:

$$\mu(x) = \frac{\ln x - \mu}{x}[1 + o(1)], \qquad x \to \infty,$$

and, therefore, $\mu(x) \to 0$ for $x \to \infty$.

In demographic literature, it is not unusual to discuss a "maximal possible age", meaning a finite number ω, for which $F(\omega) = 1$ or $F^*(\omega) = 0$, i.e. the duration of life T with certainty will not exceed ω; see, e.g., Gwinn et al. [1989b]. None of the particular distributions discussed earlier – exponential, Weibull, Gompertz, lognormal, gamma – possesses such an ω. For any of these functions $F^*(x) > 0$ for any finite x; in other words, the maximal possible age equals ∞.

The question of the existence and magnitude of ω boils down to the question of the behavior of $F(x)$ at the right-hand tail. Below in Lecture 10 we will see that accurate estimation of the distribution function $F(x)$ at the right tail, i.e. for large x, is an important problem in life insurance. Indeed, already formula (1.4) shows that the behavior of this right tail has an essential impact on the value of ET: if, for instance, $F^*(x)$ does not decrease sufficiently quickly, this expected value can well become infinite. With finite ω, the expected value ET is, certainly, finite. Although one might feel uncomfortable with the maximal possible age $\omega = \infty$, the natural and analytically convenient forms of distributions with $\omega < \infty$ are few in number.

We will, however, show one well-known distribution function with finite ω. This is the so called beta distribution function. The density of the beta distribution on the interval $[0, \omega]$ is the following function:

$$f(x) = \begin{cases} \dfrac{1}{B(a,b)} \left(\dfrac{x}{\omega}\right)^{a-1} \left(1 - \dfrac{x}{\omega}\right)^{b-1} \dfrac{1}{\omega}, & x \in [0, \omega], \\ 0, & x > \omega \text{ or } x < 0. \end{cases}$$

Here $B(a,b)$ denotes the beta function, see, e.g., Abramowitz and Stegun [1964]

$$B(a,b) = \int_0^1 t^{a-1}(1-t)^{b-1}dt, \qquad a,b > 0.$$

The beta function and gamma function are related in the following way:

$$B(a,b) = \frac{\Gamma(a)\Gamma(b)}{\Gamma(a+b)}.$$

The beta distribution function

$$F(x) = \frac{1}{B(a,b)} \int_0^x \frac{y^{a-1}(\omega - y)^{b-1}}{\omega^{a+b-1}} dy$$

can also only be expressed through elementary functions when a and b are integers. But, as in the case of gamma distribution, it is not difficult to graph $F(x)$ and its mortality rate

$$\mu(x) = \frac{x^{a-1}(\omega - x)^{b-1}}{\int_x^\omega y^{a-1}(\omega - y)^{b-1} dy}. \qquad (2.14)$$

Let us look at the asymptotic behavior of $\mu(x)$ when $x \to \omega$. Divide numerator and denominator in (2.14) by ω^{a-1}. Since

$$\left(\frac{x}{\omega}\right)^{a-1} \to 1 \quad \text{when } x \to \omega,$$

the expression we get in the numerator is asymptotically equivalent to $(\omega - x)^{b-1}$. If $a > 1$, for the integral in the denominator we will have

$$\left(\frac{x}{\omega}\right)^{a-1} \int_x^\omega (\omega - y)^{b-1} dy \leq \frac{1}{\omega^{a-1}} \int_x^\omega y^{a-1}(\omega - y)^{b-1} dy$$

$$\leq \int_x^\omega (\omega - y)^{b-1} dy,$$

because then y^{a-1} is an increasing function. If $0 < a < 1$, then y^{a-1} is a decreasing function and $\leq$ should be changed into $\geq$. It is obvious that in both cases the denominator is asymptotically equivalent to the integral

$$\int_x^\omega (\omega - y)^{b-1} \, dy = \frac{1}{b}(\omega - x)^b.$$

It therefore follows that for the beta distribution

$$\mu(x) = \frac{b}{\omega - x}[1 + o(1)], \qquad x \to \omega,$$

where ω is the maximal possible age.

As we can see, the exponent of $(\omega - x)$ always equals -1 regardless of b. This leaves us with very little choice for adequately describing data.

◇ **Exercise.** Repeat the previous exercise for the case of the beta distribution. Now you have one more parameter, and can in theory at least obtain a better fit to the data. △

◇ **Exercise.** As a simple example consider the case of the uniform distribution on the interval $[0, \omega]$. What is the corresponding failure rate $\mu(x)$? △

◇ **Exercise.** The class of distributions with regularly varying tails, see, e.g., Feller [1971], vol.2, ch. VIII.8, is often useful in many areas, such as physics and financial mathematics . A typical example is given by the so called Pareto distribution function

$$F(x) = 1 - \frac{a^\rho}{(a+x)^\rho}, \qquad x > 0,$$

where $\rho > 0$ and $a > 0$ are parameters of this distribution.What is the behavior of the corresponding failure rate for large x ? △

The empirical distribution function of duration of life

In demography, statistical problems begin where they always begin: we do not know the true distribution function F of the duration of life; but we have data, based on which we have to make inference about this unknown F, test different hypotheses which we can formulate about F, or estimate F and/or some of its parameters.

Suppose we have n independent, identically distributed durations of life $T_1, T_2, \ldots, T_n$, that is, n independent random variables, each with the same distribution function F. Let us consider a function of a random variable T and of a point x, defined as

$$I_{\{T<x\}} = \begin{cases} 1, & \text{if } T < x; \\ 0, & \text{if } T \geq x. \end{cases}$$

More generally, let A be a subset of $[0, \infty)$, say, an interval $[x_1, x_2]$, and let $I_{\{T \in A\}}$ be a function of T and A defined as

$$I_{\{T \in A\}} = \begin{cases} 1, & \text{if } T \in A; \\ 0, & \text{if } T \notin A. \end{cases} \tag{3.1}$$

The function (3.1) is called an indicator function of a set, or event, A. Indicators are very useful and will frequently be used below. In particular, consider the process in x, called binomial process in relation to (3.5),

$$z_n(x; T_1, T_2, \ldots, T_n) = \sum_{i=1}^{n} I_{\{T_i < x\}}. \tag{3.2}$$

23

For a given x this function $z_n(x; T_1, T_2, \ldots, T_n)$ is equal to the number among n individuals who have durations of life less than x. However, since $T_1, T_2, \ldots, T_n$ are random, $z_n(x; T_1, T_2, \ldots, T_n)$ is also a random variable for each x, that is, a random function in x or a random process. For any given values of $T_1, T_2, \ldots, T_n$ the trajectory of $z_n(x; T_1, T_2, \ldots, T_n)$ is a piece-wise constant, non-decreasing function of x. It equals 0 at $x = 0$ and equals n for $x > \max T_i$. It has jumps at $x = T_1, T_2, \ldots, T_n$, each of height 1.

◇ **Exercise.** Suppose we are given $n = 3$ lifetimes (in years), $T_1 = 75.5$, $T_2 = 48.6$, $T_3 = 69.1$.

Draw the graph of $z_n(x; T_1, T_2, T_3)$ in x. Is $z_n(x; T_1, T_2, T_3)$ continuous in x from the left? And from the right? If it were that $T_1 = 48.6$, $T_2 = 69.1$, $T_3 = 75.5$, would $z_n(x; T_1, T_2, T_3)$, as a function of x, change?   △

From now on we will drop $T_1, T_2, \ldots, T_n$ from the notation for z_n and write simply $z_n(x)$.

An empirical distribution function of random variables $T_1, T_2, \ldots, T_n$ is defined as

$$\widehat{F}_n(x) = \frac{1}{n} z_n(x) = \frac{1}{n} \sum_{i=1}^{n} I_{\{T_i < x\}}. \qquad (3.3)$$

From the properties of $z_n(x)$ we can easily see that $\widehat{F}_n(x)$ is a piece-wise constant, non-decreasing function of x. It equals 0 at $x = 0$ and equals 1 for $x > \max T_i$. It has jumps at $x = T_1, T_2, \ldots, T_n$ of height $1/n$ each.

Let us clarify the question of jump-points. The empirical distribution function has its first jump, from 0 to $1/n$, at the point $T_{(1)} = \min T_i$; its second jump, from $1/n$ to $2/n$, occurs at the point $T_{(2)}$, which is the second smallest duration of life; and so on until the last jump, from $(n-1)/n$ to 1, at the point $T_{(n)} = \max T_i$. The random times $T_{(1)}, T_{(2)}, \ldots, T_{(n)}$ are called order statistics (based on $T_1, T_2, \ldots, T_n$). Although we assumed that random variables $T_1, T_2, \ldots, T_n$ were independent, the order statistics can by no means be regarded as independent. For example, the distribution of $T_{(i)}$ very much depends on the values of $T_{(i-1)}$ and $T_{(i+1)}$: it can only take values between $T_{(i-1)}$ and

$T_{(i+1)}$, so even the range depends on the surrounding order statistics. Any permutation of $T_1, T_2, \ldots, T_n$ does not change the order statistics. Therefore, $\widehat{F}_n(x)$ depends not directly on $T_1, T_2, \ldots, T_n$, but only on their order statistics $T_{(1)}, T_{(2)}, \ldots, T_{(n)}$:

$$\frac{1}{n} \sum_{i=1}^{n} I_{\{T_i < x\}} = \frac{1}{n} \sum_{i=1}^{n} I_{\{T_{(i)} < x\}}. \tag{3.4}$$

Although these two sums are equal, we will see later on that they suggest quite different methods for the study of the empirical distribution function: the left side represents it as a sum of independent random functions, as each $I_{\{T_i < x\}}$ is a random function in x; while the right side represents it as a point process with stopping times $T_{(1)}, T_{(2)}, \ldots, T_{(n)}$ (cf. Lecture 8 below).

The empirical distribution function is one of the central objects in non-parametric statistics, if not in all of mathematical statistics. There is a huge body of publications on the subject and the interested reader may gain wider understanding of the theory of the empirical distribution function by reading, e.g., the monograph Shorack and Wellner [2009]. Below we will mention some classical facts concerning the asymptotic behavior of $\widehat{F}_n$, and then we shall use them.

Let us begin with the following statement:

if $T_1, T_2, \ldots, T_n$ are independent and identically distributed random variables, then for every fixed x the random variable $z_n(x)$ has the binomial distribution:

$$P\{z_n(x) = k\} = b(k; n, F(x))$$
$$= \binom{n}{k} F^k(x) [1 - F(x)]^{n-k}, \quad k = 0, 1, \ldots, n. \tag{3.5}$$

In fact, at fixed x each indicator $I_{\{T_i < x\}}$ is a Bernoulli random variable with two values, 0 and 1, and

$$P\{I_{\{T_i < x\}} = 1\} = P\{T_i < x\} = F(x).$$

Since random variables $T_1, T_2, \ldots, T_n$ are independent, then these Bernoulli random variables are also independent. Therefore $z_n(x)$ is the sum of n independent Bernoulli random variables with the probability of "success" $F(x)$, and therefore is indeed a binomial random variable with the distribution (3.5).

Once we know the distribution of the random variables $z_n(x)$ we can know the distribution of $\widehat{F}_n(x)$:

$$P\left\{\widehat{F}_n(x) = \frac{k}{n}\right\} = b(k;n,F(x)), \qquad k = 0,1,\ldots,n.$$

As a result

$$E\widehat{F}_n(x) = F(x) \quad \text{and} \quad \text{Var}\,\widehat{F}_n(x) = \frac{1}{n}F(x)\big[1-F(x)\big]. \tag{3.6}$$

⋄ **Exercise.** Find $EI_{\{T_i<x\}}$ and $\text{Var}\,I_{\{T_i<x\}}$, and then prove (3.6). △

The first equality in (3.6) shows that if random variables $T_1, T_2, \ldots, T_n$ are identically distributed with distribution function $F(x)$, then the empirical distribution function $\widehat{F}_n(x)$ is an unbiased estimator of this common $F(x)$. The second equality shows that if $T_1, T_2, \ldots, T_n$ are also independent, then the variance $\widehat{F}_n(x)$ converges to 0 at the rate of $\frac{1}{n}$, i.e. $\widehat{F}_n(x)$ is a consistent estimator in the mean-square sense. We will now prove that $\widehat{F}_n(x)$ is consistent in a much stronger sense.

First of all let us look at the following important inequality.

The exponential inequality for the binomial distribution see, e.g., (Shorack and Wellner [2009], p. 440), states that

if random variables ξ_n have the binomial distribution

$$P\{\xi_n = k\} = \binom{n}{k}p^k(1-p)^{n-k}$$

then

$$P\left\{\frac{\xi_n}{n} - p > \varepsilon\right\} \le e^{-na(p,\varepsilon)}, \tag{3.7}$$

where $a(p,\varepsilon) = p\int_0^{\varepsilon/p} \ln(1+y)\,dy$.

Consequently, if $T_1, T_2, \ldots, T_n$ are independent and identically distributed, then

$$P\{|\widehat{F}_n(x) - F(x)| > \varepsilon\} \le e^{-na(F(x),\varepsilon)} + e^{-na(F^*(x),\varepsilon)}. \tag{3.8}$$

Proofs of the inequalities (3.7) and (3.8) will be given below; but for now, let us look at how (3.8) can be applied.

For any fixed $x > 0$ and for arbitrarily small fixed $\varepsilon > 0$, values of $a(F(x), \varepsilon)$ and $a(F^*(x), \varepsilon)$ are fixed numbers. Consequently, (3.8) shows that the probability that the empirical distribution function $\widehat{F}_n(x)$ deviates from its mean $F(x)$ by more than ε decreases exponentially with increasing number of observations n. From here, in turn, it follows that

for each fixed $x > 0$

$$\widehat{F}_n(x) \to F(x), \qquad n \to \infty, \tag{3.9}$$

with probability 1.

This is a much stronger statement than the convergence of the variance $\widehat{F}_n(x)$ to 0.

Proof of (3.9). Let

$$A_n = \left\{ |\widehat{F}_n(x) - F(x)| > \varepsilon \right\}.$$

Then, according to (3.8),

$$\sum_{n=1}^{\infty} P(A_n) \leq \sum_{n=1}^{\infty} \left[e^{-na(F(x), \varepsilon)} + e^{-na(F^*(x), \varepsilon)} \right] < \infty,$$

which, using the Borel–Cantelli Lemma, implies that the probability of A_n occurring infinitely often is 0 (see, e.g., Shiryaev [1980]). In other words, the events $A_n, n = 1, 2, \ldots,$ will stop occurring and $\widehat{F}_n(x)$ will stay within an ε neighborhood of $F(x)$.

Now we come back to the exponential inequality (3.8).

Note first the inequality

$$P\{X > c\} = P\{f(X) > f(c)\} \leq \frac{Ef(X)}{f(c)},$$

true for a positive function $f(X)$ of a random variable X, which is called the Markov inequality;, see, e.g., Shiryaev [1980]. The Markov inequality is a generalization of the Chebyshev inequality, and can be proved in exactly the same way. Now note that

$$P\left\{ \frac{\xi_n}{n} - p \geq \varepsilon \right\} = P\{\xi_n - np \geq n\varepsilon\} = P\{e^{\lambda(\xi_n - np)} \geq e^{n\lambda\varepsilon}\}$$

for any positive value of the parameter λ. We will select a specific value later. Let us use the Markov inequality

$$P\{e^{\lambda(\xi_n-np)} \geq e^{n\lambda\varepsilon}\} \leq \frac{Ee^{\lambda(\xi_n-np)}}{e^{n\lambda\varepsilon}}.$$

For binomial ξ it is, however, easy to calculate the expected value $Ee^{\lambda(\xi_n-np)}$ on the right (for this it is sufficient to use the binomial formula):

$$Ee^{\lambda\xi_n} = (1+p(e^{\lambda}-1))^n.$$

Using the inequality $(1+x)^n < e^{nx}$ for this last expression we finally obtain

$$P\left\{\frac{\xi_n}{n} - p \geq \varepsilon\right\} \leq e^{np(e^{\lambda}-1)}e^{-n\lambda p - n\lambda\varepsilon} = e^{np(e^{\lambda}-1-\lambda)-n\lambda\varepsilon}.$$

Not for all values of λ does the right-hand side produce a useful inequality. Hence, it is important to choose λ that minimizes the exponent. Differentiating with respect to λ we find:

$$p(e^{\lambda}-1) = \varepsilon, \qquad \lambda = \ln\left(1+\frac{\varepsilon}{p}\right).$$

For this value of λ the exponent becomes

$$np(e^{\lambda}-1-\lambda) - n\lambda\varepsilon = n\varepsilon - n(p+\varepsilon)\ln\left(1+\frac{\varepsilon}{p}\right)$$

$$= np\left[\frac{\varepsilon}{p} - \left(1+\frac{\varepsilon}{p}\right)\ln\left(1+\frac{\varepsilon}{p}\right)\right]$$

$$= -na\left(\frac{\varepsilon}{p}\right),$$

because the expression in square brackets is equal to the integral

$$-\int_0^{\varepsilon/p} \ln(1+x)\,dx.$$

Therefore, the inequality (3.7) is proved.

Now let us show that (3.8) follows from (3.7). If random variables $T_1, T_2, \ldots, T_n$ are independent and identically distributed, then $z_n(x) =$

$n\widehat{F}_n(x)$ and $n - z_n(x) = n\widehat{F}_n^*(x)$ are both binomial random variables; and since

$$P\{|\widehat{F}_n(x) - F(x)| > \varepsilon\} = P\{\widehat{F}_n(x) - F(x) > \varepsilon\}$$
$$+ P\{\widehat{F}_n^*(x) - F^*(x) > \varepsilon\}, \qquad (3.10)$$

we can apply inequality (3.7) to both summands on the right, and this leads to (3.8).

One can prove a still stronger statement on convergence of the empirical distribution function $\widehat{F}_n(x)$ to $F(x)$ by using the so called Glivenko–Cantelli theorem (Glivenko [1933]), which is the theorem for which we were actually preparing. According to this theorem,

if $T_1, T_2, \ldots, T_n$ are independent and identically distributed with distribution function $F(x)$, then

$$\sup_x |\widehat{F}_n(x) - F(x)| \to 0, \quad as \quad n \to \infty, \qquad (3.11)$$

with probability 1.

Therefore, $\widehat{F}_n(x)$ converges to $F(x)$ not only at each particular x, but also converges uniformly in x.

◇ **Exercise.** It is well known that convergence of a sequence of functions $\varphi_n(x)$ to a limiting function $\varphi(x)$ at every x does not imply uniform convergence of φ_n to φ.

a) Construct a corresponding example with φ continuous on $[0, 1]$.
b) Nevertheless, in the case of distribution functions, point-wise convergence (3.9) implies uniform convergence (3.11), because distribution functions F have two specific properties: they are non-decreasing and bounded. Make sure that you see this without resorting to the proof below. △

Let us prove (3.11). In this proof it is not necessary for $F(x)$ to be continuous – it can be any distribution function. For a given $F(x)$ let us consider its graph

$$\Gamma_F = \{(u, x) : F(x) \leq u \leq F(x+)\}. \qquad (3.12)$$

If $F(x)$ is continuous in x, then at this x there is only one value of $F(x)$,

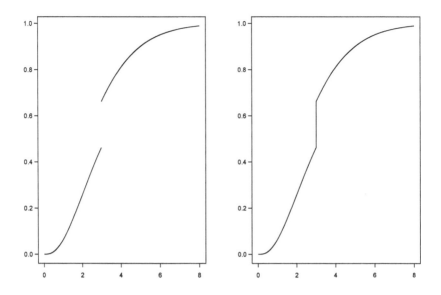

Figure 3.1 *On the left is shown a function F with one jump point, on the right, its graph Γ_F.*

$u = F(x)$; if $F(x)$ has a jump at x, then values of u fill in the whole interval from $F(x) = F(x-)$ to $F(x+)$.

Let us take an arbitrarily large integer N and consider it as fixed. Choose x_i as

$$x_i = \min\left\{x : F(x) \le \frac{i}{N} \le F(x+)\right\}, \quad i = 1,\ldots,N-1,$$

and suppose $x_0 = -\infty$ and $x_N = \infty$. With this choice of points, if $x_i < x_{i+1}$, then

$$0 \le F(x_{i+1}) - F(x_i+) \le \frac{1}{N},$$

while if $x_i = x_{i+1}$ then $F(x_{i+1}) - F(x_i) = 0$. Note also that (3.9) is true, and can be proved in exactly the same way, also for the limits from the right, $\widehat{F}_n(x+) \to F(x+)$, which is useful if F happens to be discontinuous at x. Since the number of distinct x_i does not exceed finite N, from (3.8) it follows, that

$$\max_{0 \leq i \leq N} |\widehat{F}_n(x_i) - F(x_i)| \to 0 \quad \text{and} \quad \max_{0 \leq i \leq N} |\widehat{F}_n(x_i+) - F(x_i+)| \to 0,$$

as $n \to \infty$, with probability 1.

Now for any x such that that $x_i < x < x_{i+1}$, we have

$$\widehat{F}_n(x) - F(x) \leq \widehat{F}_n(x_{i+1}) - F(x_i+) \leq \widehat{F}_n(x_{i+1}) - F(x_{i+1}) + \frac{1}{N}.$$

Similarly,

$$\widehat{F}_n(x) - F(x) \geq \widehat{F}_n(x_i+) - F(x_{i+1}) \geq \widehat{F}_n(x_i+) - F(x_i+) - \frac{1}{N}.$$

Therefore, for all such x

$$\left|\widehat{F}_n(x) - F(x)\right| \leq \max_{0 \leq i \leq N} \max \left(\left|\widehat{F}_n(x_i) - F(x_i)\right|, \left|\widehat{F}_n(x_i+) - F(x_i+)\right| \right) + \frac{1}{N}.$$

If $x = x_i$ for some i, the above inequality is still true. Since its right-hand side is independent of x, we will obtain

$$0 \leq \sup_x |\widehat{F}_n(x) - F(x)| \leq$$
$$\max_{0 \leq i \leq N} \max \left(\left|\widehat{F}_n(x_i) - F(x_i)\right|, \left|\widehat{F}_n(x_i+) - F(x_i+)\right| \right) + \frac{1}{N}.$$

The right-hand side here converges to $\frac{1}{N}$ with probability 1. Consequently,

$$0 \leq \lim_{n \to \infty} \sup_x |\widehat{F}_n(x) - F(x)| \leq \frac{1}{N}.$$

Since N can be arbitrarily large, this limit is equal to 0, which proves (3.11).

Deviation of $\widehat{F}_n(x)$ from $F(x)$ as a random process

In the previous lecture we established that the deviation of $\widehat{F}_n(x)$ from $F(x)$ converges to 0 uniformly in x. One would like, however, to find how rapid this convergence is. Let us clarify this first for just one fixed point x.

Since the variance of the difference $\widehat{F}_n(x) - F(x)$ is equal to

$$\frac{1}{n} F(x)[1 - F(x)]$$

(see (3.6)), the normalized difference $\sqrt{n}[\widehat{F}_n(x) - F(x)]$ has a "stable" variance:

$$\text{Var}\,\sqrt{n}\big[\widehat{F}_n(x) - F(x)\big] = F(x)\big[1 - F(x)\big],$$

which is in fact independent of n. But then what can we know about the behavior of this normalized difference as a random variable? That is, what can we say about the probability $P\{\sqrt{n}|\widehat{F}_n(x) - F(x)| > \lambda\}$ for any given λ?

Since

$$P\big\{\sqrt{n}|\widehat{F}_n(x) - F(x)| > \lambda\big\} = P\Big\{|\widehat{F}_n(x) - F(x)| > \frac{\lambda}{\sqrt{n}}\Big\},$$

we could have used the inequality (3.8) with $\varepsilon = \lambda/\sqrt{n}$. We have, however, the limiting expression for this probability and not just an inequality. As we have seen earlier (see (3.10)), it is sufficient to consider the one-sided probability

$$P\big\{\sqrt{n}\,[\widehat{F}_n(x) - F(x)] > \lambda\big\}.$$

As in Lecture 2, let

$$\Phi(x) = \frac{1}{\sqrt{2\pi}} \int_{-\infty}^{x} e^{-\frac{y^2}{2}} \, dy, \qquad \varphi(x) = \frac{1}{\sqrt{2\pi}} e^{-\frac{x^2}{2}}$$

denote, respectively, the standard normal distribution function and its density. According to the Berry–Esséen inequality (which we will not prove here – see, e.g., Feller [1971])

$$\sup_{-\infty < \lambda < \infty} \left| P\{\sqrt{n}[\widehat{F}_n(x) - F(x)] < \lambda\} - \Phi\left(\frac{\lambda}{\sqrt{F(x)F^*(x)}}\right) \right|$$

$$\leq \frac{F^2(x) + (F^*(x))^2}{\sqrt{nF(x)F^*(x)}}. \tag{4.1}$$

The order of magnitude of the upper bound $1/\sqrt{nF(x)F^*(x)}$ in this inequality cannot be improved uniformly for all x, and therefore on the "tails", for values of x where $F(x)$ is small or $F^*(x)$ is small, it remains unclear how well the probability

$$P\{\sqrt{n}|\widehat{F}_n(x) - F(x)| > \lambda\}$$

can be approximated by the normal distribution function. The question arises as to whether one can obtain a better approximation for the distribution function of $\sqrt{n}[\widehat{F}_n(x) - F(x)]$ for such x. This question we will consider later in Lecture 15, when we discuss the need for such approximations for certain insurance problems.

From (4.1) it follows immediately that

$$P\left\{\lambda_1 \leq \frac{\sqrt{n}[\widehat{F}_n(x) - F(x)]}{\sqrt{F(x)[1 - F(x)]}} \leq \lambda_2\right\} \rightarrow \Phi(\lambda_2) - \Phi(\lambda_1), \tag{4.2}$$

see, e.g., Cramér [1946] and Feller [1965].

As our next step consider the joint behavior of the deviations $\widehat{F}_n(x) - F(x)$ at several different x. Let $0 < x_1 < x_2 < \cdots < x_k < \infty$ be k different ages, to which we add $x_0 = 0$ and $x_{k+1} = \infty$, and consider the joint distribution function of the random variables

$$\sqrt{n}[\widehat{F}_n(x_1) - F(x_1)], \ \sqrt{n}[\widehat{F}_n(x_2) - F(x_2)], \ \ldots, \ \sqrt{n}[\widehat{F}_n(x_k) - F(x_k)]. \tag{4.3}$$

Instead of these random variables it is more convenient to consider their differences

$$\sqrt{n}[\Delta\widehat{F}_n(x_j) - \Delta F(x_j)] = \sqrt{n}\Delta[\widehat{F}_n(x_j) - F(x_j)], \quad j = 0, 1, \ldots, k, \tag{4.4}$$

where $\Delta\widehat{F}_n(x_j) = \widehat{F}_n(x_{j+1}) - \widehat{F}_n(x_j)$ is the relative frequency of deaths between ages x_j and x_{j+1}, while $\Delta F(x_j) = F(x_{j+1}) - F(x_j)$ is the probability of a death occurring between these ages. As soon as we establish the limit distribution of the random variables (4.4) it will be easy to derive the joint distribution of the random variables (4.3), which we will do in the second part of this lecture.

We start by observing that the joint distribution of the differences $\Delta z_n(x_j)$, $j = 0, 1, \ldots, k$, is multinomial:

$$P\{\Delta z_n(x_j) = l_j, \ j = 0, \ldots, k\} = \frac{n!}{\prod_{j=0}^{k} l_j!} \prod_{j=0}^{k} [\Delta F(x_j)]^{l_j}, \tag{4.5}$$

where $l_0, \ldots, l_k$ are non-negative integers, such that

$$\sum_{j=0}^{k} l_j = n.$$

Intuitively this is quite clear. Indeed, with each T_i we can associate $k + 1$ disjoint events

$$\{x_0 \le T_i < x_1\}, \ldots, \{x_k \le T_i < \infty\},$$

of which one, and only one, will take place. The probabilities of these events are $\Delta F(x_0), \ldots, \Delta F(x_k)$ and the outcomes, for different i, are independent, because $T_1, \ldots, T_n$ are independent. Then the frequencies of these events,

$$\Delta z_n(x_j) = \sum_{i=1}^{n} I_{\{x_j \le T_i < x_{j+1}\}}, \quad j = 0, 1, \ldots, k, \tag{4.6}$$

form our multinomial random vector.

◇ **Exercise.** Assuming (4.5) is true, show that the characteristic function has the form

$$E e^{i[t_0 \Delta z_n(x_0) + \cdots + t_k \Delta z_n(x_k)]}$$

$$= \sum_{\substack{l_0,\ldots,l_k: l_j \geq 0, \\ \sum_{j=0}^k l_j = n}} e^{i[t_0 l_0 + \cdots + t_k l_k]} \frac{n!}{\prod_{j=0}^k l_j!} \prod_{j=0}^k [\Delta F(x_j)]^{l_j}$$

$$= \sum_{\substack{l_0,\ldots,l_k: l_j \geq 0, \\ \sum_{j=0}^k l_j = n}} \frac{n!}{\prod_{j=0}^k l_j!} \prod_{j=0}^k \left[e^{it_j} \Delta F(x_j) \right]^{l_j} = \left[\sum_{j=0}^k e^{it_j} \Delta F(x_j) \right]^n.$$

$\triangle$

◇ **Exercise.** Prove (4.5) using the previous exercise and the following facts:

a) two distributions coincide if their characteristic functions coincide;

b) the characteristic function of the random vector $I_{\{x_j \leq T < x_{j+1}\}}$, $j = 0, \ldots, k$, equals

$$E e^{i[t_0 I_{\{x_0 \leq T < x_1\}} + \cdots + t_k I_{\{x_k \leq T < \infty\}}]} = \sum_{j=0}^k e^{it_j} \Delta F(x_j);$$

c) the characteristic function of the sum of independent random vectors is the product of the characteristic functions of the summands.

$\triangle$

Now we see that the random variables (4.4) are centered and normalized multinomial random frequencies

$$\sqrt{n} \Delta[\widehat{F}_n(x_j) - F(x_j)] = \frac{\Delta z_n(x_j) - n \Delta F(x_j)}{\sqrt{n}}, \quad j = 0, \ldots, k. \quad (4.7)$$

The frequencies $\Delta z_n(x_j)$, $j = 0, \ldots, k$, are clearly dependent random variables. They are even connected through a functional relationship:

$$\sum_{j=0}^k \Delta z_n(x_j) = n.$$

It will be useful for us to derive the covariance matrix of these frequencies:

the covariance of $\Delta z_n(x_j)$ and $\Delta z_n(x_l)$ is

$$E[\Delta z_n(x_j) - n\Delta F(x_j)][\Delta z_n(x_l) - n\Delta F(x_l)]$$
$$= n[\Delta F(x_j)\delta_{jl} - \Delta F(x_j)\Delta F(x_l)], \qquad (4.8)$$

where

$$\delta_{jl} = \begin{cases} 1, & j = l; \\ 0, & j \neq l. \end{cases}$$

Indeed, using (4.6), we can write

$$E[\Delta z_n(x_j) - n\Delta F(x_j)][\Delta z_n(x_l) - n\Delta F(x_l)]$$
$$= E\sum_{i=1}^{n}[I_{\{x_j \leq T_i < x_{j+1}\}} - \Delta F(x_j)] \cdot \sum_{i'=1}^{n}[I_{\{x_l \leq T_{i'} < x_{l+1}\}} - \Delta F(x_l)].$$

Let us multiply these sums. The expected value of the product of summands with different i and i' is 0 because T_i and $T_{i'}$ are independent:

$$E[I_{\{x_j \leq T_i < x_{j+1}\}} - \Delta F(x_j)][I_{\{x_l \leq T_{i'} < x_{l+1}\}} - \Delta F(x_l)]$$
$$= E[I_{\{x_j \leq T_i < x_{j+1}\}} - \Delta F(x_j)]E[I_{\{x_l \leq T_{i'} < x_{l+1}\}} - \Delta F(x_l)] = 0$$

for any values of j and l. Now if $i = i'$ then

$$E[I_{\{x_j \leq T_i < x_{j+1}\}} - \Delta F(x_j)][I_{\{x_l \leq T_i < x_{l+1}\}} - \Delta F(x_l)]$$
$$= EI_{\{x_j \leq T_i < x_{j+1}\}}I_{\{x_l \leq T_i < x_{l+1}\}} - \Delta F(x_j)\Delta F(x_l).$$

If $j \neq l$, the product of the indicator functions equals 0, because T_i cannot belong to two disjoint intervals at the same time; but if $j = l$ then

$$I_{\{x_j \leq T_i < x_{j+1}\}}I_{\{x_j \leq T_i < x_{j+1}\}} = I_{\{x_j \leq T_i < x_{j+1}\}}$$

and therefore

$$EI_{\{x_j \leq T_i < x_{j+1}\}}I_{\{x_j \leq T_i < x_{j+1}\}} = EI_{\{x_j \leq T_i < x_{j+1}\}} = \Delta F(x_j).$$

These last two equations imply the expression (4.8).

The random variables (4.7) are also dependent random variables and connected by the functional relationship

$$\sum_{j=0}^{k} \Delta[\widehat{F}_n(x_j) - F(x_j)] = 0. \tag{4.9}$$

From (4.8) it easily follows that

$$\mathsf{E}\sqrt{n}\Delta[\widehat{F}_n(x_j) - F(x_j)] \cdot \sqrt{n}\Delta[\widehat{F}_n(x_l) - F(x_l)]$$
$$= \Delta F(x_j)\delta_{jl} - \Delta F(x_j)\Delta F(x_l),$$

so that altogether
the covariance matrix of the random vector (4.7) has the form:

$$\mathbf{C} = \begin{pmatrix} \Delta F(x_0) & & 0 \\ & \ddots & \\ 0 & & \Delta F(x_k) \end{pmatrix} - \begin{pmatrix} \Delta F(x_0) \\ \vdots \\ \Delta F(x_k) \end{pmatrix} (\Delta F(x_0), \dots, \Delta F(x_k)).$$

$$\tag{4.10}$$

We will now establish the following fact, the central limit theorem, concerning the vector (4.7):

let us denote by $\Phi(\lambda; \mathbf{C})$ a multidimensional normal distribution with mean vector 0 and covariance matrix (4.10); here $\lambda = (\lambda_0, \dots, \lambda_k)$ is a vector in $(k+1)$-dimensional space. Then

$$\mathsf{P}\{\sqrt{n}\Delta(\widehat{F}_n(x_0) - F(x_0)) \leq \lambda_0, \dots,$$

$$\sqrt{n}\Delta(\widehat{F}_n(x_k) - F(x_k)) \leq \lambda_k\} \to \Phi(\lambda; \mathbf{C}), \quad \text{as } n \to \infty.$$
$$\tag{4.11}$$

We know that, for each n, the random variables $\sqrt{n}\Delta[\widehat{F}_n(x_j) - F(x_j)]$ are subject to the linear constraint (4.9). It is natural to expect that this will be somehow reflected in the limiting random variables. And indeed it is: the normal distribution with the distribution function $\Phi(\lambda; \mathbf{C})$ is concentrated only on the subspace of points $z_0, z_1, \dots, z_k$, which satisfy the equation $\sum_{j=0}^{k} z_j = 0$; that is, on the subspace which is orthogonal to the main diagonal, the latter defined by the vector $\mathbf{1} = (1, 1, \dots, 1)$ (with $k+1$ coordinates).

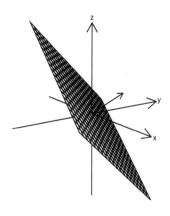

Figure 4.1 *In 3-dimensional space, $\Phi(\lambda;\mathbf{C})$ is concentrated on the chequered plane.*

◇ **Exercise.** Prove that if $X = (X_0, X_1, \ldots, X_k)$ is a Gaussian random vector with expected value 0 and covariance matrix $\mathbf{C}$, then not only $\mathrm{E}\sum_{i=0}^{k} X_i = 0$, but also $\mathrm{Var}\sum_{i=0}^{k} X_i = 0$, so that indeed $\sum_{i=0}^{k} X_i = 0$.   △

Let us now prove the limit theorem (4.11), which we will then reformulate in a vivid and more intuitive form.

This proof will be based on the fact (see, e.g., Feller [1965] or Shiryaev [1999]) that if the sequence of characteristic functions in a finite-dimensional space converges to a characteristic function, then the corresponding distribution functions converge to the limiting distribution function in every continuity point of the latter. Since the distribution function $\Phi(\lambda;\mathbf{C})$ is continuous for all λ, the convergence will occur for all λ. Recall that the characteristic function of the normal distribution function $\Phi(\lambda;\mathbf{C})$ has the form

$$\exp\left\{-\frac{1}{2}\left[\sum_{j=0}^{k}t_j^2\Delta F(x_j) - \left(\sum_{j=0}^{k}t_j\Delta F(x_j)\right)^2\right]\right\}$$

$$= \exp\left\{-\frac{1}{2}\sum_{j=0}^{k}(t_j-\bar{t})^2\Delta F(x_j)\right\}, \qquad \bar{t}=\sum_{j=0}^{k}t_j\Delta F(x_j).$$

Using the result of the previous exercise, consider now the characteristic function of the random vector (4.4):

$$\mathrm{E}\exp\left\{i\sum_{j=0}^{k}t_j\sqrt{n}\Delta[\widehat{F}_n(x_j) - F(x_j)]\right\}$$

$$= \mathrm{E}\exp\left\{i\sum_{j=0}^{k}\frac{t_j}{\sqrt{n}}\Delta z_n(x_j) - i\sqrt{n}\sum_{j=0}^{k}t_j\Delta F(x_j)\right\}$$

$$= \left[\sum_{j=0}^{k}e^{i\frac{t_j}{\sqrt{n}}\Delta F(x_j)}\right]^n \cdot \exp\left\{-i\sqrt{n}\sum_{j=0}^{k}t_j\Delta F(x_j)\right\}.$$

Taking the second term in the last expression inside the bracket we finally obtain

$$\left[\sum_{j=0}^{k}e^{i\frac{1}{\sqrt{n}}(t_j-\bar{t})}\Delta F(x_j)\right]^n. \tag{4.12}$$

Recalling the Taylor expansion up to second term, $e^z = 1+z+z^2/2+o(z^2)$, and noting that $(t_j-\bar{t})/\sqrt{n} \to 0$ when $n \to \infty$, from (4.12) one obtains

$$\left[\sum_{j=0}^{k}\left(1+i\frac{1}{\sqrt{n}}(t_j-\bar{t}) - \frac{1}{2n}(t_j-\bar{t})^2 + o\left(\frac{1}{n}\right)\right)\Delta F(x_j)\right]^n$$

$$= \left[1-\frac{1}{2n}\sum_{j=0}^{k}(t_j-\bar{t})^2\Delta F(x_j) + o\left(\frac{1}{n}\right)\right]^n$$

$$\longrightarrow e^{-\frac{1}{2}\sum_{j=0}^{k}(t_j-\bar{t})^2\Delta F(x_j)}, \qquad n \to \infty,$$

which proves (4.11).

    The central limit theorem for a multinomial random vector, which is what we have just proved (see, e.g., Cramér [1972], § 30.1), can be given a much broader and more useful interpretation by regarding the differences $\sqrt{n}\,[\widehat{F}_n(x) - F(x)]$ as a family of random variables in x, that is, as a random process. Let us introduce the following notation for this process:

$$V_n(x) = \sqrt{n}\,[\widehat{F}_n(x) - F(x)], \qquad x \geq 0. \qquad (4.13)$$

The random process (4.13) is called an empirical process.

◇ **Exercise.** Suppose $n = 3$ lifetimes, with $T_1 = 75.5$, $T_2 = 48.6$ and $T_3 = 69.1$ given. Suppose also that the distribution function of these lifetimes is $F(x) = x^2/100$, $0 \leq x \leq 100$. Draw the trajectory of $V_n(x)$, that is, the graph of $V_n(x)$ in x for given T_1, T_2 and T_3.   △

We want to show now that there exists a limiting process $V(x)$, $x \geq 0$, with a very beautiful and clear structure; and that V_n converges in a certain sense to this V.

This we will do in the next lecture.

Limit of empirical process: Brownian bridge. Distribution of χ^2 goodness of fit statistic

Let us start by introducing a generalization of Brownian motion. Let $F(x)$ be a continuous distribution function; and let $w(t)$, $0 \le t \le 1$, denote a standard Brownian motion. Then

$$W(x) = w \circ F(x) = w(F(x))$$

is called a Brownian motion with respect to time $F(x)$. In more detail, the standard Brownian motion on $[0,1]$ is a zero-mean Gaussian process with independent increments. That is, for any k and any collection of k points $0 = t_0 < t_1 < \cdots < t_k < t_{k+1} = 1$ the increments

$$\Delta w(t_j) = w(t_{j+1}) - w(t_j), \quad j = 0, \dots, k,$$

are independent Gaussian random variables. The distribution of each $\Delta w(t_j)$ has expected value 0 and variance $\Delta t_j = t_{j+1} - t_j$. Therefore, $w(0)$ is identically zero, as it has expected value 0 and variance also 0. The distribution of $w(t)$, which can also be regarded as the increment $w(t) = w(t) - w(0)$, has expected value 0 and variance

$$\mathsf{E}w^2(t) = t.$$

If now, for any given k, we choose k values of ages $0 = x_0 < x_1 < \cdots < x_k < x_{k+1} = \infty$, then the increments

$$\Delta W(x_j) = w(F(x_{j+1})) - w(F(x_j)), \quad j = 0, \dots, k,$$

are also independent Gaussian random variables with expected values 0 and variances

$$E[\Delta W(x_j)]^2 = E[w(F(x_{j+1})) - w(F(x_j))]^2 = F(x_{j+1}) - F(x_j).$$

Therefore the variance of $W(x)$,

$$EW^2(x) = F(x),$$

involves a specific distribution function, which may change from case to case. In this sense, W is not a "standard" Brownian motion anymore, but Brownian motion in time $F(x)$. It seems more natural to speak about the evolution of $W(x)$ not in terms of x, but in terms of $F(x)$.

The transformation of the time axis in something like this manner is a familiar concept. In reliability theory, for instance, the shrinking or dilating of the time axis to make the hazard rate constant produces so-called "operational time" (e.g., Gerber [1979]). Or, with modeling stopping times, it might be convenient to stretch time when stopping times tend to occur close together. Even the distribution function stretches and pulls the variable along the x axis to fit into the vertical axis in a manner reminiscent of transformation of the time axis.

The covariance functions of w and W are, respectively,

$$Ew(t)w(t') = \min(t,t'), \text{ for } t,t' \in [0,1] \qquad (5.1)$$

and

$$EW(x)W(x') = \min(F(x),F(x')) = F(\min(x,x')), \quad \text{for } x,x' \geq 0. \qquad (5.2)$$

The expected value and covariance uniquely define the distribution of a Gaussian process. So one can say that $W(x)$, $x \geq 0$, is a Gaussian process with expected value $EW(x) = 0$ and covariance function (5.2).

◇ **Exercise.** Using only (5.2), show that the variance of $\Delta W(x) = W(x+\Delta x) - W(x)$ is $\Delta F(x)$ and for any $0 \leq x_1 < x_2 < x_3$ the increments $\Delta W(x_1) = W(x_2) - W(x_1)$ and $\Delta W(x_2) = W(x_3) - W(x_2)$ are not correlated:

$$E[\Delta W(x)]^2 = \Delta F(x), \quad E\Delta W(x_1)\Delta W(x_2) = 0. \qquad (5.3)$$

In particular
$$EW(x)\Delta W(x) = 0.$$

So, W is a process with uncorrelated increments and if we also know that it is a Gaussian process, then it is a Gaussian process with independent increments; thus all of its finite-dimensional distributions are fixed. △

Now consider the process
$$V_F(x) = V(x) = W(x) - F(x)W(\infty).$$

As a linear transformation of a Gaussian process, V is itself a Gaussian process. This transformation is of quite an interesting nature: it is a projection. A linear transformation Π is called a "projection" if $\Pi^2 = \Pi$; and for $\Pi g(x) = g(x) - F(x)g(\infty)$ one can easily verify that
$$\Pi(\Pi g(x)) = \Pi g(x).$$

We will come across a similar projection later in this lecture.

The expected value of V is equal to
$$EV(x) = EW(x) - F(x)EW(\infty) = 0,$$

and its covariance function is equal to
$$\begin{aligned}
EV(x)V(x') &= EW(x)W(x') - F(x')EW(x)W(\infty) \\
&\quad - F(x)EW(\infty)W(x') + F(x)F(x')EW^2(\infty) \\
&= F(\min(x,x')) - F(x)F(x'). \tag{5.4}
\end{aligned}$$

The process $V(x)$, $x \geq 0$, is the limiting process we were looking for, as we will presently demonstrate. It is called a Brownian bridge in time $F(x)$. If
$$v(t) = w(t) - tw(1), \qquad 0 \leq t \leq 1,$$

is the standard Brownian bridge, that is, a Gaussian process with expected value $Ev(t) = 0$ and covariance
$$Ev(t)v(t') = \min(t,t') - tt',$$

then
$$V(x) = v \circ F(x) = v(F(x)).$$
As for Brownian motion $w(t)$, $v(0) = 0$; but in contrast to Brownian motion, $v(1) = 0$ as well, whence its name: it "bridges" points $(0,0)$ and $(0,1)$. Consequently $V(0) = V(\infty) = 0$.

Continuing to use the sequence of points $0 = x_0 < x_1 < \cdots < x_k < x_{k+1} = \infty$, set
$$\Delta V(x_j) = V(x_{j+1}) - V(x_j).$$
Now let us show the following:

the covariance matrix of the increments $\Delta V(x_j)$, $j = 0,\ldots,k$, is exactly the same as the covariance matrix of the increments

$$\Delta V_n(x_j) = \sqrt{n} \Delta[\widehat{F}_n(x_j) - F(x_j)], \quad j = 0,\ldots,k,$$

of the empirical process (which is the matrix $\mathbf{C}$ of (4.10)).

Indeed, using equations (5.2) and (5.3), we obtain

$$\mathsf{E}\Delta V(x_j)\Delta V(x_l)$$

$$= \mathsf{E}[\Delta W(x_j) - \Delta F(x_j)W(\infty)][\Delta W(x_l) - \Delta F(x_l)W(\infty)]$$

$$= \mathsf{E}\Delta W(x_j)\Delta W(x_l) - 2\Delta F(x_j)\Delta F(x_l) + \Delta F(x_j)\Delta F(x_l)$$

$$= \Delta F(x_j)\delta_{jl} - \Delta F(x_j)\Delta F(x_l),$$

which is the (j,l)th element of the matrix $\mathbf{C}$.

For the readers' convenience let us gather together the properties of the empirical process V_n and the Brownian bridge V, which we have obtained so far:

- If random variables $T_1, T_2, \ldots, T_n$ have the distribution function $F(x)$, then the expected values are
$$\mathsf{E}V(x) = \mathsf{E}V_n(x) = 0.$$

- Their covariance functions coincide:
$$\mathsf{E}V(x)V(x') = \mathsf{E}V_n(x)V_n(x') = F(\min(x,x')) - F(x)F(x').$$

- The covariance matrices of the increments also coincide:

$$E\Delta V(x_j)\Delta V(x_l) = E\Delta V_n(x_j)\Delta V_n(x_l) = \Delta F(x_j)\delta_{jl} - \Delta F(x_j)\Delta F(x_l).$$

- The end-point values are:

$$V(0) = V(\infty) = 0 = V_n(0) = V_n(\infty).$$

However, V is a Gaussian process, while V_n is not.

◇ **Exercise.** Using the matrix $\mathbf{C}$ of (4.10), derive the expression for $EV_n(x)V_n(x')$. We showed this expression earlier, but we did not derive it from $\mathbf{C}$. To do this, let $x = x_1 < x_2 = x'$ and $V_n(x_2) = V_n(x_1) + \Delta V_n(x_1)$. △

Now we are ready to re-formulate theorem (4.11), as well as theorem (4.2), in an intuitively clearer way. Suppose the random variable X_n has distribution function G_n, and the random variable X has distribution function G. If $G_n(x)$ converges to $G(x)$ as $n \to \infty$ at all x where $G(x)$ is continuous, then one says that G_n converges to G weakly and denotes it as $G_n \overset{w}{\to} G$. Exactly the same fact is sometimes expressed differently: one says that the random variables X_n converge in distribution to the random variable X and denote this by

$$X_n \overset{d}{\to} X.$$

The same definitions can be directly applied to random vectors and theorem (4.2) can be re-written as

$$V_n(x) \overset{d}{\to} V(x),$$

and theorem (4.11) as

$$\{\Delta V_n(x_j), \ j = 0, \ldots, k\} \overset{d}{\to} \{\Delta V(x_j), \ j = 0, \ldots, k\}. \tag{5.5}$$

At the same time, it is clear that (5.5) is equivalent to the statement

$$\{V_n(x_j), \ j = 1, \ldots, k\} \overset{d}{\to} \{V(x_j), \ j = 1, \ldots, k\}. \tag{5.6}$$

Although we do not prove this equivalence formally, intuitively it is quite clear.

The k dimensional distribution of the random vector $\{V_n(x_j),\ j = 1,\ldots k, \}$ is also called the finite-dimensional distribution of the process V_n. Since we have proved (5.5) for any collection $x_1 < \ldots < x_k$ and for any finite k, we have therefore proved the following:

if lifetimes $T_1, T_2, \ldots, T_n$ are independent and all have the same distribution function F, then all finite-dimensional distributions of the empirical process V_n converge to the corresponding finite-dimensional distributions of the Brownian bridge V.

This is usually denoted as

$$V_n \overset{d_f}{\to} V. \tag{5.7}$$

From the last theorem it follows that we know the limit distribution for any continuous function $\psi(V_n(x_j),\ j = 1,\ldots,k)$ of a finite number of $V_n(x_1), \ldots, V_n(x_k)$:

$$\psi(V_n(x_j),\ j = 1,\ldots,k) \overset{d}{\longrightarrow} \psi(V(x_j),\ j = 1,\ldots,k). \tag{5.8}$$

This is not merely a formal mathematical statement, but a fact of great practical importance. Let us demonstrate this for the case of the so-called chi-square goodness of fit statistic. This statistic has the following form:

$$\sum_{j=0}^{k} \frac{[\Delta z_n(x_j) - n\Delta F(x_j)]^2}{n\Delta F(x_j)} = \sum_{j=0}^{k} \frac{[\Delta V_n(x_j)]^2}{\Delta F(x_j)} \tag{5.9}$$

and it is, therefore, a continuous function of the values of V_n at k points. Using (5.8) we immediately obtain the limit theorem for this statistic:

$$\sum_{j=0}^{k} \frac{[\Delta V_n(x_j)]^2}{\Delta F(x_j)} \overset{d}{\longrightarrow} \sum_{j=0}^{k} \frac{[\Delta V(x_j)]^2}{\Delta F(x_j)}.$$

We will now show that the sum of squares on the right side has a remarkable distribution function — the chi-square distribution function with k degrees of freedom:

$$\chi_k^2(x) = \frac{1}{2^{k/2}\Gamma(k/2)} \int_0^x y^{\frac{k}{2}-1} e^{-\frac{y}{2}}\, dy.$$

Hence, this limit distribution does not depend on F and does not depend on the selection of the boundary points x_i.

To show this fact it is easiest to invoke a geometric argument, similar to the one illustrated in Figure 4.1. We present this argument in several steps.

Before we do this, note that the chi-square distribution function $\chi_k^2(x)$ is the distribution function of the sum of squares of k independent Gaussian random variables, each with expected value 0 and variance 1. Here, however, we have the sum of $k+1$ and not k summands; and the random variables

$$Y_j = \frac{\Delta V(x_j)}{\sqrt{\Delta F(x_j)}}, \quad j = 0, \dots, k, \tag{5.10}$$

are Gaussian, but not independent and no longer with variance 1.

First, from the expression for the covariance matrix $\mathbf{C}$ in (4.10) it follows that the covariance matrix $\mathbf{D}$ of the weighted increments Y_j has the form

$$\mathbf{D} = \begin{pmatrix} 1 & & 0 \\ & \ddots & \\ 0 & & 1 \end{pmatrix} - \begin{pmatrix} \sqrt{\Delta F(x_0)} \\ \vdots \\ \sqrt{\Delta F(x_k)} \end{pmatrix} \left(\sqrt{\Delta F(x_0)}, \dots, \sqrt{\Delta F(x_k)} \right).$$

Let us rewrite it as

$$\mathbf{D} = I - \sqrt{\Delta F} \sqrt{\Delta F}^T,$$

where I denotes the $(k+1) \times (k+1)$ identity matrix and $\sqrt{\Delta F}$ is slightly awkward but clear notation for the $(k+1)$-dimensional vector

$$\sqrt{\Delta F} = \left(\sqrt{\Delta F(x_0)}, \dots, \sqrt{\Delta F(x_k)} \right)^T.$$

Next, note that the vector $\sqrt{\Delta F}$ has a convenient property: for any F and whatever choice of $x_1, \dots, x_k$, its Euclidean norm is always 1:

$$\| \sqrt{\Delta F} \|^2 = \sqrt{\Delta F}^T \sqrt{\Delta F} = \sum_{j=0}^{k} \Delta F(x_j) = 1.$$

From this it follows that the matrix $\mathbf{D}$ is idempotent, or is a projection:

DD = **D**. More importantly, let $Z = (Z_0, \ldots, Z_k)^T$ denote a vector of independent Gaussian random variables, each with the expected value 0 and variance 1; then the vector (5.10) itself can be represented as a projection of Z:

$$Y_j = Z_j - \left[\sum_{l=0}^{k} Z_l \sqrt{\Delta F(x_l)}\right] \sqrt{\Delta F(x_j)}, \quad j = 0, \ldots, k,$$

or, in matrix notation,

$$Y = Z - \left[Z^T \sqrt{\Delta F}\right] \sqrt{\Delta F}.$$

The covariance matrix of Z is, obviously, I; and the covariance matrix of the right-hand side is **D**, which proves the equality.

Further, note that the sum $[Z^T \sqrt{\Delta F}]$ has a Gaussian distribution with standard parameters:

$$E\left[\sum_{l=0}^{k} Z_l \sqrt{\Delta F(x_l)}\right] = \left[\sum_{l=0}^{k} E Z_l \sqrt{\Delta F(x_l)}\right] = 0$$

and

$$\text{Var}\left[\sum_{l=0}^{k} Z_l \sqrt{\Delta F(x_l)}\right] = \sum_{l=0}^{k} \Delta F(x_l) = 1,$$

and that this sum and all coordinates of Y are independent:

$$E\left(Z_j - \left[\sum_{l=0}^{k} Z_l \sqrt{\Delta F(x_l)}\right] \sqrt{\Delta F(x_j)}\right)\left[\sum_{l=0}^{k} Z_l \sqrt{\Delta F(x_l)}\right]$$

$$= E Z_j \left[\sum_{l=0}^{k} Z_l \sqrt{\Delta F(x_l)}\right] - E\left[\sum_{l=0}^{k} Z_l \sqrt{\Delta F(x_l)}\right]^2 \sqrt{\Delta F(x_j)}$$

$$= \sqrt{\Delta F(x_j)} - \sqrt{\Delta F(x_j)} = 0.$$

Therefore, the vector Y and the random variable $[Z^T \sqrt{\Delta F}]$ are independent.

It only remains to note that the vectors Y and $\sqrt{\Delta F}$ are orthogonal to each other:

$$Y^T \sqrt{\Delta F} = \sum_{l=0}^{k} \frac{\Delta V(x_l)}{\sqrt{\Delta F(x_l)}} \sqrt{\Delta F(x_l)}$$

$$= \sum_{l=0}^{k} \Delta V(x_l) = V(\infty) = 0.$$

But then,

$$Z = Y + \left[Z^T \sqrt{\Delta F} \right] \sqrt{\Delta F}$$

represents a random vector Z as a sum of two orthogonal (random) vectors, and one can use Pythagoras' theorem to deduce that

$$Z^T Z = Y^T Y + \left[Z^T \sqrt{\Delta F} \right]^2 \sqrt{\Delta F}^T \sqrt{\Delta F},$$

or

$$Z^T Z = Y^T Y + \left[Z^T \sqrt{\Delta F} \right]^2. \tag{5.11}$$

The quadratic form

$$Z^T Z = \sum_{j=0}^{k} Z_j^2$$

has a chi-square distribution function χ_{k+1}^2 with $k+1$ degrees of freedom; the summands on the right-hand side are independent and the distribution of $[Z^T \sqrt{\Delta F}]^2$ is χ_1^2. So, the distribution of $Y^T Y$ must be chi-square with k degrees of freedom χ_k^2 (for this "must be" see the exercise below).

Thus we have proved that

the limit distribution function of the chi-square goodness of fit statistic (5.9), as $n \to \infty$, is $\chi_k^2(x)$.

◇ **Exercise.** Can one take

$$Z_j = \frac{\Delta W(x_j)}{\sqrt{\Delta F(x_j)}}$$

in the proof above? △

◇ **Exercise.** To prove rigorously that $Y^T Y$ cannot have any other than the χ_k^2 distribution function, reason as follows: use the fact that the characteristic function of χ_k^2, or equivalently the characteristic function of a chi-square random variable, is

$$\hat{\chi}_k^2(t) = \left(\frac{1+i2t}{1+4t^2}\right)^{k/2}.$$

Therefore, the characteristic function of $Z^T Z$ is

$$\hat{\chi}_{k+1}^2(t) = \left(\frac{1+i2t}{1+4t^2}\right)^{(k+1)/2},$$

while the characteristic function of $[Z^T \sqrt{\Delta F}]^2$ is

$$\hat{\chi}_1^2(t) = \left(\frac{1+i2t}{1+4t^2}\right)^{1/2}.$$

Let $\phi(t)$ denote the characteristic function of $Y^T Y$. Since the summands in (5.11) are independent, the characteristic function of the sum is the product of the characteristic functions, and we have:

$$\left(\frac{1+i2t}{1+4t^2}\right)^{(k+1)/2} = \phi(t)\left(\frac{1+i2t}{1+4t^2}\right)^{1/2}.$$

Therefore

$$\phi(t) = \hat{\chi}_k^2(t).$$

If you agree with this, you have done the exercise. $\triangle$

There are, however, statistics that depend not on a finite number but on an infinite number of values of $V_n(x)$. One example is the Kolmogorov–Smirnov statistic

$$\sqrt{n}\sup_{x\geq 0}|\widehat{F}_n(x) - F(x)| = \sup_{x\geq 0}|V_n(x)|$$

or its one-sided version

$$\sqrt{n}\sup_{x\geq 0}[\widehat{F}_n(x) - F(x)] = \sup_{x\geq 0}V_n(x).$$

It appears that for quite a large number of such statistics a statement similar to (5.8) is again valid:

$$\psi\{V_n(x), x \geq 0\} \xrightarrow{d} \psi\{V(x), x \geq 0\}. \tag{5.12}$$

In particular,

$$\sup_{x \geq 0} |V_n(x)| \xrightarrow{d} \sup_{x \geq 0} |V(x)|. \tag{5.13}$$

The distribution function of the statistic on the right is Kolmogorov's distribution function (see, for example, Kolmogorov [1992], Shorack and Wellner [2009]):

$$K(\lambda) = 1 - 2\sum_{k=1}^{\infty} (-1)^k e^{-2k^2\lambda^2}. \tag{5.14}$$

Tables for this distribution and comments on its use and various approximations can be found in Bol'shev and Smirnov [1983]. And so, (5.13) is the limit theorem for the Kolmogorov–Smirnov statistic:

$$P\left\{\sup_{x \geq 0} |V_n(x)| < \lambda\right\} \to K(\lambda), \qquad n \to \infty.$$

Extension of the finite-dimensional convergence into the "infinite-dimensional" convergence (5.12) is denoted as

$$V_n \xrightarrow{d} V, \qquad n \to \infty.$$

The precise description of the class of statistics ψ, for which (5.12) is valid, is the subject of the theory of convergence of random processes in distribution (see, e.g., Billingsley [1977]). The theory for stochastic processes with more complicated "time" is discussed, for example, in van der Vaart and Wellner [1996].

Lecture 6

Statistical consequences of what we have learned so far. Two-sample problems

As we have already mentioned at the beginning of Lecture 3, we are developing our theory in the following way: the duration of life has a distribution function F, which is used to determine all the probabilities in which we are interested. This distribution function, however, is not known; so we build statistical estimates of F based on the available data, such as the empirical distribution function $\widehat{F}_n$, or other functions to be discussed later. For example, for the two ages $x < x'$ the difference $F(x') - F(x)$ is the probability $\mathsf{P}\{x \leq T < x'\}$,

$$\mathsf{P}\{x \leq T < x'\} = F(x') - F(x),$$

while the difference $\widehat{F}_n(x') - \widehat{F}_n(x)$ is only its statistical estimate. Then we study the properties of our estimate: how accurate it is, how quickly it converges to F, the probability of deviations from F of a certain magnitude and so on. We may also take a somewhat different approach: Based on general considerations we can choose a hypothetical distribution function F; and then, comparing it to our estimate, we can decide how well our hypothesised function F corresponds to the empirical data.

Quite often, however, in demography textbooks estimates like $\widehat{F}_n$, obtained through empirical data, are interpreted as a distribution function. Quantities such as $\widehat{F}_n(x') - \widehat{F}_n(x)$ or $\widehat{F}_n^*(x) = 1 - \widehat{F}_n(x)$ are simply called probabilities, and used as such. The function F, the nub of the theory, is ignored.

This confounding of theory and empiricism occurs so frequently, that one suspects systematic reasons for such an error. The following analogy may help to explain one possible cause. The earth's surface has been formed as a result of so many variable forces over so long a period, that a given landscape can be considered as random, one of many possible versions that could have resulted from the past. Nevertheless, we have only one earth, and earth scientists naturally restrict their scope to the particular geography that we happen to have.

In like vein, one can say that demography studies a given specific population and it is not deemed necessary to consider the mechanism giving rise to many "similar" populations. Perhaps that is why observed frequencies are regarded as something absolute and referred to simply as probabilities.

This seemingly pragmatic and healthy point of view, we think, is incorrect, and is subject to a number of objections. Let us discuss some of them.

A. Any given specific population of humans will change considerably and become a different population, even over as short a time period as 10–15 years. If anything were to remain unchanged over time, it would be the general, abstract statistical characteristics of the population – and this is primarily the distribution function F.

B. Everyone would agree that the empirical distribution function or other statistical characteristics become more "accurate" the larger the number of observations used. For example, $\widehat{F}_{2n}$ is more accurate than $\widehat{F}_n$, it is closer to something absolute, something true. But what is that absolute if not F?

C. Let us compare duration of life in two different populations, say in Sydney and Melbourne, or in Sydney now and 50 years ago. Suppose we want to establish whether the probability of death in any particular age group is different in these populations. But frequencies of death in different age groups in these two populations would almost certainly be different – how could it be possible for observations to be exactly the same in two different populations? These differences might be small, or not so small, but how could we form an opinion on what is small and what is not without understanding that $\widehat{F}_n$ naturally fluctuates around F; and without knowing how big these natural fluctuations are likely to be? Once we distinguish the distribution function F from its estimates such as $\widehat{F}_n$, we obtain natural, clear, and elegant methods for comparison of different populations.

And so, throughout this course of lectures, distribution functions and their statistical estimates will be different objects.

Consider also another question: why is it that we interpret observed durations of life as particular realizations of independent and identically distributed random variables $T_1, \ldots, T_n$?

The assumption of independence seems quite acceptable in a very wide range of problems. One can talk about genetically conditioned correlation between the durations of life of close relatives, incidences of massive loss of life in groups of people (e.g., during catastrophes), and so on, but in reasonably large populations possible dependences become diluted and do not play an essential role.

As to the assumption that lifetimes are identically distributed, this assumption is quite often not satisfied. For example, the distribution function of duration of life is known to differ between males and females; and insurance companies assume that smokers and non-smokers have different distributions of the duration of life.

Distributions of lifetimes may also vary across socio-economic groups and for people belonging to different generations. The distribution of the duration of life for males born in Europe in 1920, for instance, is quite different from the distribution for those born in Europe in 1970, if only because of the former living through the Second World War.

The set of people with the same distribution of duration of life is called in demography a cohort. For example, a set of people of the same gender, born in the same period of time, and the same place, and belonging to the same socio-economic group is usually considered as a cohort.

Given observations on lifetimes of a cohort of size n, we now consider how one can carry out statistical inference about the underlying F.

Estimation of F. Denote by $T_1, \ldots, T_n$ the observed lifetimes of n individuals and let the corresponding empirical distribution function

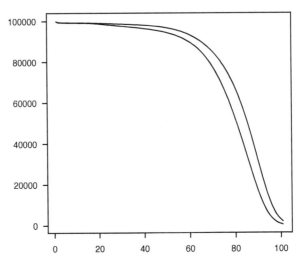

Figure 6.1 *Instead of probability of survival until a certain age by an individual, it is common in demography to speak about the number of survivals per conventional 100 000 of a cohort. In our notations it is simply* $100\,000\,(1 - \widehat{F}_n(x))$. *The figure shows the graphs of this function for the male (solid line) and female population of New Zealand in 2004.*

be, as previously,

$$\widehat{F}_n(x) = \frac{1}{n} \sum_{i=1}^{n} I_{\{T_i < x\}}.$$

Let us use $\widehat{F}_n$ to estimate the unknown F. What have we learned about the properties of this estimate? If these n individuals belong to a cohort, i.e. if $T_1, \ldots, T_n$ are independent and identically distributed random variables, then (see (3.6)):

$$\mathsf{E}\widehat{F}_n(x) = F(x)$$

for all x. In statistical terminology this means that $\widehat{F}_n$ is an unbiased estimator of F. We also know that

$$\sup_x |\widehat{F}_n(x) - F(x)| \to 0 \quad \text{with probability 1,} \quad n \to \infty,$$

see (3.11), which in statistical terms means that $\widehat{F}_n$ is strongly consistent. Since $\sqrt{n}\sup_x |\widehat{F}_n(x) - F(x)|$ is now "stable" and does not converge to 0, the rate of convergence of $\sup_x |\widehat{F}_n(x) - F(x)|$ to 0 is $1/\sqrt{n}$.

Moreover we know the probability that $\widehat{F}_n(x)$ will not deviate from $F(x)$ by more than $\lambda/\sqrt{n}$ for any λ,

$$P\left\{\sup_x |\widehat{F}_n(x) - F(x)| \le \frac{\lambda}{\sqrt{n}}\right\} = K_n(\lambda).$$

These probabilities are well tabulated (see, e.g., Shorack and Wellner [2009], ch.8), and we have a number of computer programs to calculate these and similar probabilities to high accuracy and for very large n (see Lecture 15).

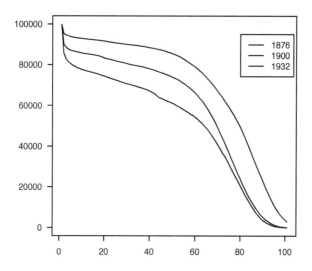

Figure 6.2 *Number of survivals until age x in a conventional cohort of 100 000 persons, for three cohorts: male population of New Zealand born in 1876, 1900 and 1932; see Dunstan et al. [2006]. The graphs show a strong improvement in mortality on the national level over just 30 years: child mortality strongly decreased and longevity of adults increased.*

Using probabilities $K_n(\lambda)$ and their limit $K(\lambda)$ one can take a further step and construct confidence limits for F. Indeed,

$$K_n(\lambda) = P\left\{\sup_x |\widehat{F}_n(x) - F(x)| \le \frac{\lambda}{\sqrt{n}}\right\}$$

$$= P\left\{\widehat{F}_n(x) - \frac{\lambda}{\sqrt{n}} \le F(x) \le \widehat{F}_n(x) + \frac{\lambda}{\sqrt{n}} \text{ for all } x\right\}$$

$$= P\left\{\max\left(\widehat{F}_n(x) - \frac{\lambda}{\sqrt{n}}, 0\right) \leq F(x) \leq \min\left(\widehat{F}_n(x) + \frac{\lambda}{\sqrt{n}}, 1\right) \text{ for all } x\right\}.$$

The last equality is true because $0 \leq F(x) \leq 1$. Therefore the random corridor or "strip" with the boundaries for each x

$$\left[\max\left(\widehat{F}_n(x) - \frac{\lambda}{\sqrt{n}}, 0\right), \min\left(\widehat{F}_n(x) + \frac{\lambda}{\sqrt{n}}, 1\right)\right] \tag{6.1}$$

covers the distribution function F with probability $K_n(\lambda)$. This is the desired confidence strip for F. If one chooses λ so that $K_n(\lambda) = 1 - \alpha$ with the given α or $K(\lambda) = 1 - \alpha$, then F will be covered by this confidence strip with probability $1 - \alpha$, or probability asymptotically equal to $1 - \alpha$.

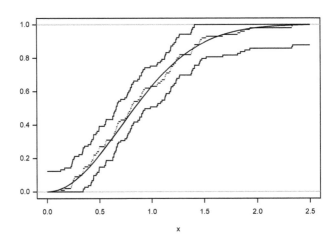

Figure 6.3 *Distribution function F (smooth line) and its estimator $\widehat{F}_n$ along with the boundaries (6.1). The sample size here is $n = 100$ and $\alpha = 0.05$.*

Testing hypotheses about F. From *a priori* considerations, theoretical motives or experience, one can advance an assumption that the plausible distribution function of our $T_1, \ldots, T_n$ is a given distribution function F_0. The statement

$$F = F_0 \tag{6.2}$$

remains, however, just that: a statistical hypothesis. This hypothesis may be true or not and needs to be tested. How does one do this?

Note, first, that we did not formulate an alternative hypothesis – if not F_0, then what? Or rather we have very broad alternative

$$F \neq F_0,$$

which, in other words, means that we would like to detect any deviations of F from F_0. For this purpose we use the so-called goodness of fit tests. The Kolmogorov–Smirnov test discussed next is one such test.

Let us centre the empirical distribution function at the hypothetical F_0; that is, let us consider

$$\sqrt{n}\,[\widehat{F}_n(x) - F_0(x)]. \tag{6.3}$$

If the hypothesis is true, then this process is the empirical process $V_n(x)$, considered in Lecture 4, and therefore the Kolmogorov–Smirnov statistic

$$\sqrt{n}\,\sup_x |\widehat{F}_n(x) - F_0(x)|$$

has the distribution function $K_n(\lambda)$. This however is true only if the hypothesis (6.2) is true. In order to emphasise this, let us introduce F_0 into our notation on the left side:

$$\mathsf{P}\left\{\sqrt{n}\,\sup_x |\widehat{F}_n(x) - F_0(x)| < \lambda \mid F_0\right\} = K_n(\lambda).$$

If the hypothesis (6.2) is not true, then the process (6.3) will acquire non-zero shift, i.e. its expected value will be the function $\sqrt{n}\,[F(x) - F_0(x)]$ and not 0. It is known from the theory of goodness of fit tests that for every $F \neq F_0$, the statistic

$$\sqrt{n}\,\sup_x |\widehat{F}_n(x) - F_0(x)| = \sqrt{n}\,\sup_x |\widehat{F}_n(x) - F(x) + F(x) - F_0(x)|$$

becomes stochastically greater than under the null hypothesis $F = F_0$; i.e., for all λ

$$\mathsf{P}\{\sqrt{n}\,\sup_x |\widehat{F}_n(x) - F_0(x)| > \lambda \mid F_0\}$$
$$\leq \mathsf{P}\{\sqrt{n}\,\sup_x |\widehat{F}_n(x) - F_0(x)| > \lambda \mid F\}.$$

Let us choose λ so that $K_n(\lambda) = 1 - \alpha$ for as small a given α as we may wish to choose (for example, $\alpha = 0.10$, $\alpha = 0.05$, $\alpha = 0.001$). If under these circumstances data are such that the discrepancy between $\widehat{F}_n$ and F_0 is greater than $\lambda/\sqrt{n}$, i.e. if

$$\sqrt{n} \sup_x |\widehat{F}_n(x) - F_0(x)| > \lambda, \tag{6.4}$$

then under the null hypothesis this could have happened only with small probability α; whereas under the alternative hypothesis (6.4) is more likely. Thus, according to the Kolmogorov–Smirnov test, if (6.4) is true, then the null hypothesis (6.2) is rejected at significance level α. Otherwise, the null hypothesis is accepted.

We will not discuss the theory of Kolmogorov–Smirnov tests in greater detail, and neither do we discuss the theory of other goodness of fit tests. We simply note that the basic framework remains unchanged for these other tests. For instance, to use the chi-square goodness of fit test one chooses the number of classes (or groups) $k+1$ (say $k+1 = 3$ or 5 or 7 or 10, and we do not recommend a higher number), and boundaries for classes $0 < x_1 < \cdots < x_k < \infty$, so that $n\Delta F_0(x_j) \geq 10$ for each j. Then one can calculate the chi-square statistic

$$\sum_{j=0}^{k} \frac{[\Delta z_n(x_j) - n\Delta F_0(x_j)]^2}{n\Delta F_0(x_j)} \tag{6.5}$$

which should be compared with (5.9). If the hypothesis (6.2) is true, then the distribution of the statistic (6.5), as we already know, should be close to the chi-square distribution with k degrees of freedom. If the hypothesis is not true, then the statistic (6.5) again becomes stochastically greater:

$$P\left\{ \sum_{j=0}^{k} \frac{[\Delta z_n(x_j) - n\Delta F_0(x_j)]^2}{n\Delta F_0(x_j)} > \lambda \,\Big|\, F_0 \right\}$$

$$\leq P\left\{ \sum_{j=0}^{k} \frac{[\Delta z_n(x_j) - n\Delta F_0(x_j)]^2}{n\Delta F_0(x_j)} > \lambda \,\Big|\, F \right\} \tag{6.6}$$

for all λ (and sufficiently large n). Again, choosing λ so that $\chi_k^2(\lambda) =$

$1-\alpha$ for sufficiently small α, we will reject the null hypothesis $F = F_0$ if the value of the statistic (6.5) exceeds this λ: for such an event, we emphasise, will not only have a small probability (of only α) under the null hypothesis but also a higher probability under the alternatives.

Testing of equality of distribution functions in two different cohorts. Consider $T_{11}, T_{12}, \ldots, T_{1n}$ observations for a cohort of n individuals and let F_1 be the distribution function of each T_{1i}. Similarly, suppose we have $T_{21}, T_{22}, \ldots, T_{2m}$ independent observations of another cohort of m individuals with distribution function F_2. Let us see how one can test the hypothesis

$$F_1 = F_2. \tag{6.7}$$

The problem of testing this hypothesis, the so-called two-sample problem, is the mathematical representation of the heuristic question: "Have the conditions of life for the cohorts been similar?" or, say, of the question: "Was the difference in conditions of life for the two cohorts reflected in their mortality?"

Note that in testing the hypothesis (6.7) we do not ask what the common distribution function is; we are simply asking whether the two functions F_1 and F_2 are identical.

Let us consider the so-called two-sample empirical process:

$$V_{n,m}(x) = \sqrt{\frac{nm}{n+m}} \left[\widehat{F}_{1n}(x) - \widehat{F}_{2m}(x) \right],$$

where, obviously, $\widehat{F}_{1n}$ and $\widehat{F}_{2m}$ denote the corresponding empirical distribution functions of the first and second cohorts/samples. If the hypothesis (6.7) is true, then $E\widehat{F}_{1n}(x) = E\widehat{F}_{2m}(x)$, so that

$$EV_{n,m}(x) = 0;$$

while under the alternative hypothesis $F_1 \neq F_2$, the expectation

$$EV_{n,m}(x) = \sqrt{\frac{nm}{n+m}} \left[F_1(x) - F_2(x) \right]$$

can be a function quite different from 0. Under the null hypothesis the correlation function of $V_{n,m}$ is equal to

$$EV_{n,m}(x)V_{n,m}(x') = F(\min(x,x')) - F(x)F(x'), \tag{6.8}$$

where $F \, (= F_1 = F_2)$ is the common hypothetical distribution function. Note that (6.8) coincides with the correlation function of the Brownian bridge (5.4).

◇ **Exercise.** Derive the correlation function (6.8) by representing $V_{n,m}$ as a weighted difference of two independent empirical processes

$$V_{n,m}(x) = \sqrt{\frac{m}{n+m}} \, \sqrt{n} \left[\widehat{F}_{1n}(x) - F(x) \right]$$
$$- \sqrt{\frac{n}{n+m}} \, \sqrt{m} \left[\widehat{F}_{2m}(x) - F(x) \right].$$

△

The distribution function of the two-sample Kolmogorov–Smirnov statistic

$$P\left\{ \sup_x |V_{n,m}(x)| < \lambda \mid F_1 = F_2 \right\} \tag{6.9}$$

has been extensively tabulated for various n and m and is included in statistical packages. For n and $m \geq 100$, and even for smaller n and m, it is possible to reliably use the distribution function $K(\lambda)$, which is the limit of (6.9) (see (5.14)). This statement, as well as the corresponding statement for the two-sample chi-square statistic, follows from the general limit theorem for the two-sample empirical process:
   if the null hypothesis (6.7) is true, then

$$V_{n,m} \overset{d}{\to} V. \tag{6.10}$$

◇ **Exercise.** Show that the limit theorem (6.10) is true by using the limit theorem

$$V_{1n} \overset{d}{\to} V_1, \quad n \to \infty \quad (\text{and} \quad V_{2m} \overset{d}{\to} V_2, \quad m \to \infty),$$

the independence of V_{1n} and V_{2m}, and the representation of $V_{n,m}$ as in the previous exercise.

△

The careful reader may have already formed a question in his mind: why is it that we do not discuss the dependence of probabilities

$$P\left\{\sqrt{n}\sup_{x}|\widehat{F}_n(x) - F_0(x)| > \lambda \mid F_0\right\}$$

on F_0, or the dependence of similar probabilities (6.9) on the hypothetical common distribution function; and what happened to the assumption of continuity of F made in Lecture 5? The answer is that we should indeed have re-stated the continuity assumption for F in the estimation problem, for F_0 in the testing problem and for F_1 and F_2 in the two-sample problem; and that under these assumptions of continuity the distributions of our statistics do not depend on F, or F_0, or the common distribution function in the two-sample problem.

This is the fundamental property of the classical goodness of fit test. It is based on a lemma, given in Kolmogorov [1933] (see Kolmogorov [1992]; a brief history of the development that followed from this lemma is given in Khmaladze [1992]). The lemma states that

if a random variable T has a continuous distribution function F, then the random variable $F(T) = U$ has the uniform distribution on $[0, 1]$.

Indeed

$$P\{U \leq t\} = P\{F(T) \leq t\} = P\{T \leq F^{-1}(t)\} = F(F^{-1}(t)) = t.$$

In the next exercise, the reader is asked to verify that the following result (6.11) is a consequence of the lemma above;

if F is continuous, then the process

$$v_n(t) = V_n(F^{-1}(t)), \tag{6.11}$$

derived from the empirical process V_n using the time transformation $t = F(x)$, is an empirical process based on independent uniformly distributed random variables $U_1 = F(T_1), U_2 = F(T_2), \ldots, U_n = F(T_n)$, in which each U_j is uniformly distributed on $[0,1]$.

Therefore v_n has a standard distribution, independent of F: the construction of v_n involves F, but its distribution is free of this F.

The process v_n is called the uniform empirical process. The existence of v_n as a distinct object of substantial mathematical interest was emphasised in Doob [1949].

◇ **Exercise.**

a) Show that the statement (6.11) about v_n is true. To do this verify that

$$\widehat{F}_{nU}(t) = \widehat{F}_n(F^{-1}(t)) = \frac{1}{n}\sum_{i=1}^{n} I_{\{T_i \leq F^{-1}(t)\}}$$

is indeed the empirical distribution function based on the sequence $U_1, U_2, \ldots, U_n$.

b) Verify that

$$\sup_{0 \leq x < \infty} |V_n(x)| = \sup_{0 \leq t \leq 1} |v_n(t)|.$$

c) More generally, verify that if the statistic $\psi[V_n, F]$, based on V_n and possibly on F, is such that

$$\psi[V_n, F] = \varphi[v_n],$$

where φ depends only on v_n, then the distribution of the statistic $\psi[V_n, F]$ is free of F (i.e., the distribution of $\psi[V_n, F]$ does not depend on F).   △

Testing parametric hypotheses. Unexpected example – exponentiality of durations of rule of Roman emperors

The problem discussed in the previous lecture of testing whether two samples come from the same distribution arises frequently in practice. But the problem of testing the hypothesis $F = F_0$, where F_0 is a completely defined distribution function, is in practice quite rare. Usually the hypothetical distribution function contains parameters with values not fixed under the hypothesis.

One can, for example, have a hypothesis that the distribution function of lifetimes is the Weibull distribution function. But it is dependent on two parameters λ and k in (2.3), and we do not have any prior grounds for choosing values for these hypothetical parameters. On the contrary – their values under the hypothesis remain free and for a given sample have to be estimated from the data. The same is true for all the other distribution functions introduced in Lecture 2: the Gompertz distribution (2.5), the logistic distribution (2.8), the Γ distribution (2.9) and the lognormal distribution (2.13) – they all depend on parameters that remain unknown under the hypothesis and have to be estimated from the sample.

Instead of a fixed hypothetical distribution function F_0, then, we have a parametric family of distribution functions $\mathbb{F} = \{F_\theta(x), \ \theta \in \mathbb{R}^d\}$, dependent on a d-dimensional parameter θ, and we want to test the parametric hypothesis

$$F \in \mathbb{F},$$

i.e. show that there exists such a value θ_0 of the parameter θ that $F = F_{\theta_0}$.

Since we do not know the true value of θ, we cannot use the empirical process

$$V_n(x, \theta_0) = \sqrt{n}\left[\widehat{F}_n(x) - F_{\theta_0}(x)\right],$$

and statistics based on it. Instead we have to use the so called parametric empirical process

$$V_n(x, \hat{\theta}_n) = \sqrt{n}\left[\widehat{F}_n(x) - F_{\hat{\theta}_n}(x)\right],$$

where $\hat{\theta}_n$ denotes the estimate of the parameter obtained from the sample $T_1, \ldots, T_n$. It is clear that a statistician desires consistent estimates: i.e., $\hat{\theta}_n \to \theta_0$ in probability as $n \to \infty$, so that the distribution function of $T_1, \ldots, T_n$ is indeed $F_{\theta_0}(x)$.

Because $\hat{\theta}_n$ is close to θ_0, it is not clear at first glance how significant the difference is between the processes $V_n(x, \hat{\theta}_n)$ and $V_n(x, \theta_0)$.

To understand why, in fact, a significant difference does exist, let us note that although $V_n(x, \theta)$ can be a very smooth function of θ, it is also a function of n. However, $\hat{\theta}_n \to \theta_0$ at the same time as $n \to \infty$; and as a function of the two variables θ and n, $V_n(x, \theta)$ is not continuous.

To look at this in greater detail, suppose that the null hypothesis is true and that θ_0 is indeed the true value. Let us expand $F_{\hat{\theta}_n}(x)$ as a function of $\hat{\theta}_n$ around θ_0 as far as the linear term, assuming the necessary differentiability property:

$$V_n(x, \hat{\theta}_n) = V_n(x, \theta_0) - \frac{\partial}{\partial \theta}F_{\theta_0}(x)\sqrt{n}(\hat{\theta}_n - \theta_0) + o_P\left(\sqrt{n}(\hat{\theta}_n - \theta_n)\right).$$

Although $\hat{\theta}_n - \theta_0 \to 0$ in probability, in the overwhelming majority of cases the normalized deviation $\sqrt{n}(\hat{\theta}_n - \theta_0)$ does not converge to 0 at all, but converges in distribution to a random variable (so that the remainder term converges in probability to 0). Thus, for $n \to \infty$, the difference between $V_n(x, \hat{\theta}_n)$ and $V_n(x, \theta_0)$ is not asymptotically small, and the substitution of an estimate changes the characteristics of the empirical process. In particular, even when after the time transformation $t = F_{\theta_0}(x)$ the process $V_n(x, \theta_0)$ becomes distribution free (i.e. free of F_{θ_0}), the process $V_n(x, \hat{\theta}_n)$ is no longer distribution free: neither for $t = F_{\theta_0}(x)$, nor for $t = F_{\hat{\theta}_n}(x)$.

As a result, statistics $\psi\left[V_n(\cdot, \hat{\theta}_n), F_{\hat{\theta}_n}(\cdot)\right]$ of the type (c) in the last exercise of Lecture 6 are neither distribution free, nor asymptotically distribution free. In particular, the distribution of the Kolmogorov–Smirnov statistic

$$\widehat{D}_n = \sup\left|V_n(x, \hat{\theta}_n)\right| \tag{7.1}$$

will no longer converge to the Kolmogorov distribution. Its limiting distribution, like distributions of other statistics, will be dependent on the family $\mathbb{F}$, on the choice of the estimate $\hat{\theta}_n$ and, generally speaking, on the value of θ_0 in the same family of distributions.

Let us briefly list some references on this problem: the change of properties of the parametric empirical process $V_n(x, \hat{\theta}_n)$ compared with the empirical process $V_n(x, \theta_0)$ was discovered in Gikhman [1954] and Kac et al. [1955]; a limit theorem on convergence of $V_n(\cdot, \hat{\theta}_n)$ to a certain Gaussian process was proved relatively late in Durbin [1973]; and the structure of the limiting process $\widehat{V}$ was studied in Khmaladze [1979]. The problem of not being distribution free, described earlier, has also been studied for individual statistics; in the case of the chi-square statistic see Moore [1971], Nikulin [1973] and Gvanceladze and Chibisov [1978]. The same problem in a different setting was explored in Nikabadze and Stute [1977] and Koenker and Xiao [2004]. Finally Koul and Swordson [2010] give more references on the subject.

We can discuss a number of specific ways to avoid the aforementioned difficulty, but a relatively complete solution of the problem is based on the following approach: one needs to adopt an alternative point of view of the construction of the empirical process and center $\widehat{F}_n(x)$ somewhat differently. In this way we obtain another version of the parametric empirical process, which after the time transformation $t = F_{\hat{\theta}_n}(x)$ again becomes asymptotically distribution free; and consequently, so do the statistics based on it.

The general theory of these transformed processes is given in Khmaladze [1981] and Khmaladze [1993]. The bibliography on this

subject is nowadays quite extensive. Computer programs for numerical calculations in **R** are included in the software package described in
Koenker [2005]. A course of lectures on applications to demography
is not the place for delving more formally into technical aspects of this
theory. Instead we shall explain its main idea through a particular, and
somewhat unexpected, example of testing a parametric hypothesis and
show how to construct the transformed process in this case.

7.1   Durations of rule of Roman emperors

Duration of human life, as well as duration of life of other biological
organisms and lifetimes of most mechanisms, cannot follow the exponential distribution: organisms are ageing, mechanisms are wearing
out, and the failure rate $\mu(x)$ increases. Indeed, the force of mortality in
human populations has little to do with the constant function $\mu(x) = \lambda$
of the exponential distribution (see Lecture 2). Although the exponential distribution is used in the insurance business, usually it is only in
the context of durations of life of those who died in accidents, natural
disasters, and such.

However, while preparing for a lecture in demography and looking
for interesting data for an example, we came across durations of the
reigns of Roman emperors of the period usually designated by historians as the "decline and fall" of the Roman Empire. It was strange to
see that, as a "statistical ensemble" of data, these durations reveal good
agreement with a general form of the exponential distribution function.

If, as we will discover below, statistical analysis confirms this
agreement, it implies that reigns were terminated in the same "unpredictable" and "purely random" way, without any "preparation" – more
or less as accidents happen or α-particles are emitted during radioactive disintegration, which we discussed in Lecture 2.

It would seem much more natural to expect that, overall, political reigns also "age" and are terminated as a result of accumulated
difficulties of a political, economic, social or personal nature. But an
exponential pattern of the data contradicts this point of view.

One could say that an emperor's life was subject to much danger
– wars, political intrigues and the like; and his rule could therefore
cease in a way analogous to that of an accident. However, we are not

convinced on that point: if the dangers were high, the supreme power of the period would certainly possess enough means to defend itself from them.

For a more complete discussion and comparisons with some European Monarchies, as well as with data on Chinese emperors' reigns, we refer to Khmaladze et al. [2007] and Khmaladze et al. [2010], to which we add only one methodological remark.

The very idea of applying formal statistical methods to data as complex as durations of rule for such a long-lived and extensive organism as the Roman Empire may seem totally absurd. And indeed, in an analysis of this sort the statistician should act with great care and circumspection. This point notwithstanding, however, we recall that in other cases an analogous approach has led to remarkably deep and interesting discoveries. For example, a completely inadequate treatment of great literary works as a "statistical ensemble" of words led, nonetheless, to discovery of such laws of word usage as Zipf's law, or Karlin–Rouault's law (see Zipf [1949] and Khmaladze [2001]) and gave rise to the development of the whole field of textual research (see Baayen [2002]).

Now consider the data.

As an attachment to this lecture we show the chronology of rule of Roman emperors as it is given in Kienast [1990]. It covers a longer period than that which historians would usually call the "decline and fall" period, i.e. it starts with Augustus Octavian (26 BC – 14 AD) and not, say, with Nerva (96 – 98 AD). The last reign included in our sample is the reign of Theodosius I, the last Emperor before the empire was divided into the Western and Eastern Roman Empires. (This and two other chronologies were considered in Khmaladze et al. [2007].)

To the reader not immersed in the minutiae of Roman history, it may look unexpected that there were sometimes two or even three emperors at the same time, dividing power or competing for it.

For example, the periods of reign of Valerian (August 253 – July 260) and Gallienus (October 253 – September 268), and those of Maxentius (October 306 – October 312) and Constantin I (July 306 – May 337), have considerable overlap. Also overlapping are the reigns of Emperors Constans (September 337 – January 350), Constantin II

(September 337 – April 342) and Constantius II (September 337 – November 361).

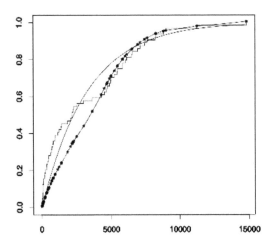

Figure 7.1 *Empirical distribution function $\widehat{F}_n$ of the durations of reigns (in days) according to Kienast [1990] and its approximation by the exponential distribution function (solid line). The solid line with asterisks is the graph of the compensator K_n. For the composition of K_n and asymptotic behaviour of the difference $\widehat{F}_n - K_n$ see the next section.*

Figure 7.1 shows the empirical distribution function of the durations of reigns and its approximation by the exponential distribution function. Visually the fit is good. The value of the statistic (7.1) equals $0.152 \times \sqrt{53} = 1.1$, which gives a probability value of 18% for the Kolmogorov distribution, comfortably away from the 10%, 5% and 1% significance levels conventionally used for rejection of a null hypothesis. This is however not so informative, because the limiting distribution of the statistic $\hat{D}_n$ is not the Kolmogorov distribution.

7.2 Testing exponentiality

To derive a new version of the empirical process let us consider an increment of the empirical distribution function

$$\Delta \widehat{F}_n(t) = \widehat{F}_n(t + \Delta t) - \widehat{F}_n(t).$$

Its unconditional expected value under the hypothesis of exponentiality is $\Delta F_\lambda(t) = e^{-\lambda t} - e^{-\lambda(t + \Delta t)}$, into which we have to substitute the estimate $\widehat{\lambda}_n$ instead of λ. However, we suggest using the conditional expected value of $\Delta \widehat{F}_n(t)$, given its "past" until the moment t. The essential point is that we will include not only all values of $\widehat{F}_n(s), s \leq t$, but also the estimator $\widehat{\lambda}_n$ in this past. In doing this we assume that we are using the maximum likelihood estimator $\widehat{\lambda}_n = (\frac{1}{n} \sum_{i=1}^{n} T_i)^{-1}$. Note that the average in the brackets can be written as the integral

$$\frac{1}{n} \sum_{i=1}^{n} T_i = \int_0^\infty x \, d\widehat{F}_n(x).$$

Therefore, we are talking about the conditional expected value

$$\mathsf{E}\left[\Delta \widehat{F}_n(t) \,\big|\, \widehat{F}_n(s), s \leq t, \widehat{\lambda}_n\right] = \mathsf{E}\left[\Delta \widehat{F}_n(t) \,\bigg|\, \widehat{F}_n(s), s \leq t, \int_0^\infty x \, d\widehat{F}_n(x)\right].$$

If we know the trajectory of $\widehat{F}_n$ up to the moment t, then we also know the part $\int_0^t x \, d\widehat{F}_n(x)$ of the whole integral. Thus, our conditional expected value reduces to

$$\mathsf{E}\left[\Delta \widehat{F}_n(t) \,\bigg|\, \widehat{F}_n(s), s \leq t, \int_t^\infty x \, d\widehat{F}_n(x)\right].$$

This expression can be simplified if we accept, without proof, that conditioning on $\{\widehat{F}_n(s), s \leq t\}$ can be replaced by conditioning on $\widehat{F}_n(t)$; that is, given the value of $\widehat{F}_n(t)$, we accept that the evolution of the trajectory up to this t does not have any influence on the expected future evolution. Let us also agree to use, instead of the conditional expected value

$$\mathsf{E}\left[\Delta \widehat{F}_n(t) \,\bigg|\, \widehat{F}_n(t), \int_t^\infty x \, d\widehat{F}_n(x)\right],$$

simply the linear regression of the increment $\Delta\widehat{F}_n(t)$ on $\widehat{F}_n(t)$ and $\int_t^\infty x d\widehat{F}_n(x)$.

To obtain this linear regression, let us rearrange the random variables in the condition to obtain uncorrelated random variables. This is not a necessary step, but a convenient one in order to facilitate the derivation.

Namely, replace $\int_t^\infty x d\widehat{F}_n(x)$ by

$$\int_t^\infty \left(x - t - \frac{1}{\lambda}\right) d\widehat{F}_n(x) = \frac{1}{n}\sum_{i=1}^n \left(T_i - \mathsf{E}\left[T_i|T_i > t\right]\right) I_{\{T_i \geq t\}}. \quad (7.2)$$

Conditioning on $\widehat{F}_n(t)$ and $\int_t^\infty x d\widehat{F}_n(x)$ is, certainly, equivalent to conditioning on $1 - \widehat{F}_n(t)$ and

$$\int_t^\infty \left(x - t - \frac{1}{\lambda}\right) d\widehat{F}_n(x) = \int_t^\infty x d\widehat{F}_n(x) - \left(t + \frac{1}{\lambda}\right)\left(1 - \widehat{F}_n(t)\right), \quad (7.3)$$

but, for every t, the latter two random variables are uncorrelated.

◇ **Exercise.** Show that if a random variable T has the exponential distribution function $1 - e^{-\lambda x}$, then

$$\mathsf{E}[T|T \geq t] = t + \frac{1}{\lambda},$$

cf. (7.2); and

$$Var(T - \mathsf{E}[T|T > t]|T \geq t) = \frac{1}{\lambda^2}.$$

The best way to do this is to recall the conditional distribution function of $T - t$ given $T \geq t$. △

Now recall that linear regression of a random variable X on uncorrelated random variables Y and Z is a linear combination $aY + bZ$, such that the difference $X - aY - bZ$ is not correlated either with Y or Z. For this to be true it must be that

$$a = \frac{\mathsf{E}X(Y - \mathsf{E}Y)}{VarY}, \quad b = \frac{\mathsf{E}X(Z - \mathsf{E}Z)}{VarZ}.$$

Applying this to our situation with $Y = 1 - \widehat{F}_n(t)$ and Z equal to (7.2) we see that

$$\mathsf{E}Y = 1 - F_\lambda(t) \quad \text{and} \quad \mathsf{E}Z = 0,$$

and that

$$VarY = \frac{1}{n}F_\lambda(t)(1 - F_\lambda(t)),$$

while

$$VarZ = \frac{1}{n}Var\left(T_i - \mathsf{E}\left[T_i | T_i > t\right] \middle| T_i \geq t\right)\mathsf{P}\left(T_i \geq t\right)$$

$$= \frac{1}{n}\frac{1}{\lambda^2}\left(1 - F_\lambda(t)\right).$$

Therefore, for the regressand $\Delta\widehat{F}_n(t)$, the coefficients a and b are

$$a = -n\frac{\mathsf{E}\Delta\widehat{F}_n(t)\left[\widehat{F}_n(t) - F_\lambda(t)\right]}{F_\lambda(t)\left(1 - F_\lambda(t)\right)},$$

$$b = n\lambda^2\frac{\mathsf{E}\Delta\widehat{F}_n(t)\int_t^\infty \left(x - t - \frac{1}{\lambda}\right)d\widehat{F}_n(x)}{1 - F_\lambda(t)}.$$

From Lecture 4 – see the display formula just before (4.10) – it is easy to see that

$$\mathsf{E}\Delta\widehat{F}_n(t)\left[\widehat{F}_n(t) - F_\lambda(t)\right] = -\frac{1}{n}\Delta F_\lambda(t)F_\lambda(t),$$

and one can accept without proof (or do the following exercise) that

$$\mathsf{E}\Delta\widehat{F}_n(t)\int_t^\infty \left(x - t - \frac{1}{\lambda}\right)d\widehat{F}_n(x) = -\frac{1}{n\lambda}\Delta F_\lambda(t) + o(\Delta t).$$

Thus, dropping negligible terms,

$$a = \frac{\Delta F_\lambda(t)}{1 - F_\lambda(t)}, \quad b = -\lambda\frac{\Delta F_\lambda(t)}{1 - F_\lambda(t)}, \tag{7.4}$$

and the desired linear regression of $\Delta\widehat{F}_n(t)$ on the uncorrelated random variables $1 - \widehat{F}_n(t)$ and (7.3) reduces to

$$\frac{\Delta F_\lambda(t)}{1 - F_\lambda(t)}\left[1 - \widehat{F}_n(t)\right] - \lambda\frac{\Delta F_\lambda(t)}{1 - F_\lambda(t)}\int_t^\infty \left(x - t - \frac{1}{\lambda}\right)d\widehat{F}_n(x).$$

Noticing that for small Δt

$$\frac{\Delta F_\lambda(t)}{1 - F_\lambda(t)} \approx \mu(t)\Delta t = \lambda \Delta t$$

we can now say that the difference

$$\Delta \widehat{F}_n(t) - \lambda \left(1 - \widehat{F}_n(t)\right)\Delta t + \lambda^2 \int_t^\infty \left(x - t - \frac{1}{\lambda}\right) d\widehat{F}_n(x)\Delta t \qquad (7.5)$$

is uncorrelated with the past of $\widehat{F}_n(s), s \leq t$, and $\hat{\lambda}_n$; reverting to integral form,

$$W_n(t) = \sqrt{n}\left[\widehat{F}_n(t) - \right.$$

$$\left. \lambda \int_0^t \left(1 - \widehat{F}_n(s)\right)ds + \lambda^2 \int_0^t \int_s^\infty \left(x - s - \frac{1}{\lambda}\right)d\widehat{F}_n(x)ds\right]$$

is a process with uncorrelated increments.

The process

$$K_n(t) = \lambda \int_0^t \left(1 - \widehat{F}_n(s)\right)ds - \lambda^2 \int_0^t \int_s^\infty \left(x - s - \frac{1}{\lambda}\right)d\widehat{F}_n(x)ds$$

can be called a compensator of $\widehat{F}_n$ – it "compensates" it, or adjusts it, to a process with uncorrelated increments. We shall treat compensators more generally in Lectures 8 and 9 below.

◇ **Exercise.** Consider a partition of the positive real-line $\{t \geq 0\}$ by points $0 = t_0 < t_{1m} < t_{2m} < \cdots < t_{km} < \cdots$ and consider the increments (7.5) over these points, so that $\Delta t_{km} = t_{k+1,m} - t_{km}$. Show that partial sums of these increments have a limit when $\max_k \Delta t_{km} \to 0$, as $m \to \infty$. This limit is the process W_n. △

◇ **Exercise.** Show that W_n can be written as the following linear transformation of V_n:

$$W_n(t) = V_n(t) + \lambda \int_0^t V_n(s)ds + \lambda^2 \int_0^t \int_s^\infty \left(x - s - \frac{1}{\lambda}\right)dV_n(x)ds$$

$$(7.6)$$

where, we recall, $V_n(t) = \sqrt{n}[\widehat{F}_n(t) - F_\lambda(t)]$ is the empirical process.   △

It is obvious that K_n depends on λ. However, we have now remarkable flexibility with substituting any consistent estimator $\hat{\lambda}$, not necessarily the maximum likelihood estimator, and not even necessarily a $\sqrt{n}$-consistent estimator:

suppose $\sqrt{n}(\hat{\lambda}_n - \lambda)^2$ is small in probability:

$$\sqrt{n}(\hat{\lambda}_n - \lambda)^2 = o_P(1), \quad as \ n \to \infty;$$

then if we replace V_n in (7.6) by

$$\hat{V}_n(t) = \sqrt{n}[\hat{F}_n(t) - F_{\hat{\lambda}_n}(t)],$$

the change in W_n will be asymptotically negligible.

Indeed, consider a Taylor expansion in λ up to the second term:

$$\sqrt{n}\left(F_{\hat{\lambda}_n}(t) - F_\lambda(t)\right) = \frac{\partial}{\partial\lambda}F_\lambda(t)\sqrt{n}(\hat{\lambda}_n - \lambda) + R_n(t)\sqrt{n}\frac{(\hat{\lambda}_n - \lambda)^2}{2},$$

where the reminder term can be written as

$$R_n(t) = \frac{\partial^2}{\partial\lambda^2}F_{\tilde{\lambda}}(t) = -t^2 e^{-\tilde{\lambda}t}$$

and $\tilde{\lambda}$ is a point between $\hat{\lambda}_n$ and λ. It is not difficult to verify that not only

$$F_\lambda(t) + \lambda\int_0^t F_\lambda(s)ds + \lambda^2 \int_0^t \int_s^\infty \left(x - s - \frac{1}{\lambda}\right)dF_\lambda(x)ds = 0,$$

as we realized in the last exercise, but also that this transformation applied to $\frac{\partial}{\partial\lambda}F_\lambda(t)$ is identically equal to 0:

$$\frac{\partial}{\partial\lambda}F_\lambda(t) + \lambda\int_0^t \frac{\partial}{\partial\lambda}F_\lambda(s)ds + \lambda^2 \int_0^t \int_s^\infty \left(x - s - \frac{1}{\lambda}\right)d\frac{\partial}{\partial\lambda}F_\lambda(x)ds = 0.$$

Therefore,

$$\hat{V}_n(t) + \lambda\int_0^t \hat{V}_n(s)ds + \lambda^2 \int_0^t \int_s^\infty \left(x - s - \frac{1}{\lambda}\right)d\hat{V}_n(x)ds$$
$$= W_n(t) - \rho_n(t)\sqrt{n}(\hat{\lambda}_n - \lambda)^2,$$

where ρ_n is just the transformation of R_n:

$$\rho_n(t) = R_n(t) + \lambda \int_0^t R_n(s)ds + \lambda^2 \int_0^t \int_s^\infty \left(x - s - \frac{1}{\lambda}\right) dR_n(x)ds.$$

With the R_n as above and $\tilde{\lambda} \to \lambda$ in probability, it is clear that $\rho_n(t)$ remains a bounded function in t uniformly in n, so that

$$\sup_t |\rho_n(t)| \sqrt{n}(\hat{\lambda}_n - \lambda)^2 \to 0, \quad n \to \infty,$$

in probability, which proves the statement.

For a similar result for a general parametric family see, e.g., Khmaladze [1981] or Koul and Swordson [2010].

Being a process with uncorrelated increments, in the limit as $n \to \infty$, W_n becomes a Gaussian process with uncorrelated and hence independent increments, such that

$$\mathsf{E}W(t) = 0, \qquad \mathsf{E}W(t)W(t') = F_{\lambda_0}(\min\{t, t'\}).$$

This is the process which we called in Lecture 5 a Brownian motion with respect to time $F_{\lambda_0}(t)$. Therefore, after a time transformation $\tau = F_{\lambda_0}(t)$ it becomes a standard Brownian motion:

$$w(\tau) = W(t), \qquad \tau = F_{\lambda_0}(t),$$

with

$$\mathsf{E}w(\tau) = 0 \quad \text{and} \quad \mathsf{E}w(\tau)w(\tau') = \min\{\tau, \tau'\}.$$

The analogue of w for finite n is the process

$$w_n(\tau) = W_n(F_{\lambda_0}^{-1}(\tau)).$$

The process w_n is derived from W_n in the same way that the uniform empirical process v_n was derived from V_n; see (6.11) . It is more or less clear that a wide class of statistics of the form $\psi[W_n, F_{\hat{\lambda}_n}]$, which may depend on $F_{\hat{\lambda}_n}$ as well but are such that

$$\psi\left[W_n, F_{\hat{\lambda}_n}\right] = \varphi[w_n],$$

where φ depends only on w_n, have as limit distribution the distribution of $\varphi[w]$. As was demonstrated in Haywood and Khmaladze [2008], the rate of convergence of w_n to w is quite good.

The limit distribution of the Komogorov–Smirnov statistic

$$\lim_{n\to\infty} P(\sup_t |W_n(t)| < z) = P(\sup_\tau |w(\tau)| < z)$$

is well known. It can be found, e.g., in Feller [1965], and tables of the distribution are given at the end of the lecture. The limit distribution function of the omega-square statistic

$$\lim_{n\to\infty} P(\int_0^\infty W_n(t)^2 dF_{\hat{\lambda}_n}(t) < z) = P(\int_0^1 w(\tau)^2 d\tau < z)$$

is also well known; see Martynov [1978]. The shorter version of the tables is given at the end of the lecture.

The continuous line with crosses in Figure 7.1 is the graph of the compensator K_n for durations of reign. The value of the Kolmogorov–Smirnov statistic has the value

$$\sqrt{n}\sup_t |\widehat{F}_n(t) - K_n(t)| = \sup_t |W_n(t)| = 1.52,$$

which corresponds to a probability value of 26%. The agreement with the hypothesis of exponentiality was very good.

◇ **Exercise.** For any realization of random variables $X_1,\ldots,X_n$ the empirical distribution function $\widehat{F}_n$ is just a discrete distribution function, concentrated on these n values and attaching weight $\frac{1}{n}$ to each of them. The integral $\int g(x)\,d\widehat{F}_n(x)$ is, therefore, no different in its form from the expected value of $g(\cdot)$ with respect to a discrete distribution:

$$\int g(x)\,d\widehat{F}_n(x) = \frac{1}{n}\sum_{i=1}^n g(X_i).$$

a) Show that

$$\mathrm{Cov}\left(\int g_1(x)\,d\widehat{F}_n(x),\ \int g_2(x)\,d\widehat{F}_n(x)\right)$$

$$= \frac{1}{n} \left(\int g_1(x) g_2(x) \, dF(x) - \int g_1(x) \, dF(x) \int g_2(x) \, dF(x) \right).$$

b) Use a) to show that the covariance of

$$1 - \widehat{F}_n(t) \quad \text{and} \quad \int_t^\infty \left(x - t - \frac{1}{\lambda} \right) d\widehat{F}_n(x)$$

is 0. $\qquad\qquad\qquad\qquad\qquad\qquad\qquad\qquad\qquad\qquad\qquad\qquad\qquad\triangle$

To conclude this lecture we remark that basically the same approach is applicable to a very wide class of parametric families of distribution functions $\{F_\theta(x), \ \theta \in \mathbb{R}^d\}$. In the general case, instead of the integral

$$\int_t^\infty x \, d\widehat{F}_n(x)$$

one should use the integral

$$\int_t^\infty \psi_f(x) d\widehat{F}_n(x),$$

where $\psi_f(x)$ is what was called until the 1960s the Fisher informant, but has in more recent times been labelled as the score function,

$$\psi_f(x) = \frac{\partial}{\partial \theta} \ln f_\theta(x).$$

This integral then can be centered, which will again make the random variables

$$1 - \widehat{F}_n(t) \quad \text{and} \quad \int_t^\infty \left(\psi_f(x) - \mathsf{E}\Big[\psi_f(T) \big| T > t \Big] \right) d\widehat{F}_n(x),$$

uncorrelated. It can be shown that

$$\mathsf{E}\Delta\widehat{F}_n(t) \int_t^\infty \left(\psi_f(x) - \mathsf{E}\Big[\psi_f(T) \big| T > t \Big] \right) d\widehat{F}_n(x)$$

$$= \Delta\widehat{F}_\theta(t) \left(\psi_f(t) - \mathsf{E}\Big[\psi_f(T) \big| T > t \Big] \right) + o(\Delta t).$$

Consequently, the coefficients a and b of (7.4) should be replaced by

$$a = \frac{\Delta F_\lambda(t)}{1 - F_\lambda(t)}$$

and

$$b = \frac{\Delta F_\lambda(t)}{1 - F_\lambda(t)} \frac{\psi_f(t) - E\left[\psi_f(T)|T > t\right]}{Var\left(\psi_f(T) - E\left[\psi_f(T)|T > t\right]\middle| T > t\right)}.$$

The rest of the development above remains unchanged.

We will not enter into more details of how the latter variance can be effectively calculated and what to do in the case of a vector parameter θ – see, e.g., Koenker [2005] and Koul and Swordson [2010].

7.3*   Chronology of reign of Roman emperors from Kienast (1990)

Augustus	(16. Jan. 27 v. Chr. - 19. Aug. 14 n. Chr.)
Tiberius	(19. Aug. 14 - 16. Marz 37)
Caligula	(18. Marz 37-24. Jan. 41)
Claudius	(24. Jan. 41 - 13. Okt. 54)
Nero	(13. Okt. 54 - 9. Juni 68)
Galba	(8. Juni 68 - 15. Jan. 69)
Otho	(15. Jan. - 16. April 69)
Vitellius	(2. Jan. - 20 Dez. 69)
Vespasian	(1. Juli 69 - 23 Juni 79)
Titus	(24. Juni 79 - 13. Sept. 81)
Domitian	(14. Sept. 81 - 18. Sept. 96)
Nerva	(18. Sept. 96 - 27.[?] Jan. 98)
Trajan	(28. Jan. 98 - 7. Aug. 117)
Hadrian	(11. Aug. 117 - 10. Juli 138)
Antoninus Pius	(10. Juli 138 - 7. Marz 161)
Mark Aurel	(7. Marz 161 - 17. Marz 180)
Commodus	(17. Marz 180 - 31. Dez. 192)
Pertinax	(31. Dez. 192 - 28. Marz 193)
Didius Iulianus	(28. Marz - 1. Juni 193)
Septimius Severus	(9. April 193 - 4. Febr. 211)
Caracalla	(4. Febr. 211 - 8. April 217)
Macrinus	(11. April 217 - 8. Juni 218)
Elagabal	(16. Mai 218 - 11. Marz 222)
Severus Alexander	(13. Marz 222 - Febr./Marz 235)
Maximinus Thrax	(Febr./Marz 235 - Mitte April [?] 238)
Gordian I.	(Jan [?] 238)
Gordian II.	(Jan [?] 238)
Pupienius	(Ende Jan./Anf. Febr. [?] - Anf. Mai [?] 238)
Balbinus	(Jan./Febr. [?] - Mai [?] 238)
Gordian III.	(Jan./Febr. [?] 238 - Anf. 244)
Philippus Arabs	(Anf. 244 - Sept./Okt. 249)
Decius	(Sept./Okt. 249 - Juni 251)
Trebonianus Gallus	(Juni [?] 251 - Aug. [?] 253)
Aemilius Aemilianus	(Juli/Aug. - Sept./Okt. 253)
Valerian	(Juni/Aug. 253 - Juni [?] 260)

Gallienus	(Sept./Okt. 253 - ca. Sept. 268)
Claudius II. Gothicus	(Sept./Okt. 268 - Sept. 270)
Quintillus	(September 270)
Aurelian	(Sept. 270 - Sept./Okt. 275)
Tacitus	(Ende 275 - Mitte 276)
Florianus	(Mitte - Herbst 276)
Probus	(Sommer 276 - Herbst 282)
Carus	(Aug./Sept. 282 - Juli/Aug. 283)
Numerianus	(Juli/Aug. [?] 283 - Nov. 284)
Carinus	(Fruhjahr 283 - Aug./Sept. 285)
Diocletian	(20. Nov. 284 - 1. Mai 305)
Maximian	(Okt./Dez. 285 - ca. Juli 310)
Constantius I.	(1. Marz 293 - 15. Juli 306)
Galerius	(21 Mai [?] 293 - Anf. MAi 311)
Maximinus Daia	(1. Mai 305 - Spatsommer 313)
Severus II.	(1. Mai 305 - Marz/April 307)
Maxentius	(28. Okt. 306 - 28. Okt. 312)
Licinius	(11. Nov 307 - 19 Sept. 324)
Constantin I.	(25. Juli 306 - 22. Mai 337)
Constantin II.	(9. Sept. 337 - Anf. April 340)
Constans	(9. Sept. 337 - 18. Jan. 350)
Constantius II.	(9. Sept. 337 - 3. Nov. 361)
Julian	(ca. Febr. 360 - 26./27. Juni 363)
Jovian	(27. Juni 363 - 17. Febr. 364)
Valentinian I.	(25. Febr. 364 - 17. Nov. 375)
Valens	(28. Marz 364 - 9. Aug. 378)
Gratian	(24. Aug. 367 - 25. Aug. 383)
Valentinian II.	(22. Nov. 375 - 15. Mai 392)
Theodosius I.	(19. Jan. 379 - 17. Jan. 395)

7.4*   Distributions of Kolmogorov–Smirnov and ω^2 statistics from a standard Brownian motion

x	Pr	x	Pr
0.1	0.0000	2.5	0.9752
0.2	0.0000	2.6	0.9814
0.3	0.0000	2.7	0.9861
0.4	0.0006	2.8	0.9898
0.5	0.0092	2.80703	0.9900
0.6	0.0414	2.9	0.9925
0.7	0.1027	3.0	0.9946
0.8	0.1852	3.1	0.9961
0.9	0.2776	3.2	0.9973
1.0	0.3708	3.3	0.9981
1.1	0.4593	3.4	0.9987
1.2	0.5404	3.5	0.9991
1.3	0.6130	3.6	0.9994
1.4	0.6770	3.7	0.9996
1.5	0.7328	3.8	0.9997
1.6	0.7808	3.9	0.9998
1.7	0.8217	4.0	0.9999
1.8	0.8563	4.1	0.9999
1.9	0.8851	4.2	0.9999
1.95996	0.9000	4.3	1.0000
2.0	0.9090	4.4	1.0000
2.1	0.9285	4.5	1.0000
2.2	0.9444	4.6	1.0000
2.24140	0.9500	4.7	1.0000
2.3	0.9571	4.8	1.0000
2.4	0.9672	4.9	1.0000
2.49771	0.9750	5.0	1.0000

Table 7.1 *Values of distribution function of Komogorov–Smirnov statistic* $\sup_t |w(t)|$ *from Brownian motion, calculated by R. Brownrigg. Wider tabels are available at* http://www.msor.vuw.ac.nz/ ~ray

$x \setminus \Delta x$	0.00	0.02	0.04	0.06	0.08
0.0	0.0000	0.0006	0.0176	0.0583	0.1090
0.1	0.1610	0.2106	0.2566	0.2988	0.3374
0.2	0.3727	0.4051	0.4348	0.4622	0.4875
0.3	0.5110	0.5328	0.5532	0.5723	0.5901
0.4	0.6069	0.6227	0.6377	0.6518	0.6651
0.5	0.6778	0.6899	0.7013	0.7122	0.7226
0.6	0.7325	0.7420	0.7511	0.7597	0.7680
0.7	0.7760	0.7836	0.7909	0.7979	0.8047
0.8	0.8111	0.8174	0.8233	0.8291	0.8347
0.9	0.8400	0.8451	0.8501	0.8549	0.8595
1.0	0.8639	0.8682	0.8723	0.8763	0.8802
1.1	0.8839	0.8875	0.8909	0.8943	0.8975
1.2	0.9007	0.9037	0.9066	0.9094	0.9122
1.3	0.9148	0.9174	0.9198	0.9222	0.9246
1.4	0.9268	0.9290	0.9311	0.9331	0.9351
1.5	0.9370	0.9389	0.9407	0.9424	0.9441
1.6	0.9457	0.9473	0.9488	0.9503	0.9518
1.7	0.9532	0.9545	0.9558	0.9571	0.9583
1.8	0.9595	0.9607	0.9618	0.9629	0.9640
1.9	0.9650	0.9660	0.9670	0.9680	0.9688
$x \setminus \Delta x$	0.0	0.2	0.4	0.6	0.8
2	0.9697	0.9772	0.9828	0.9870	0.9902
3	0.9925	0.9943	0.9957	0.9967	0.9975
4	0.9981	0.9985	0.9989	0.9991	0.9993

Table 7.2 *Values of distribution function of ω^2-statistic $\int_0^t w^2(t)dt$ from Brownian motion, calculated for this lecture by G. Martynov*

p	0.5	0.9	0.95	0.975	0.99	0.995
x_p	0. 2905	1.1958	1.6557	2.1347	2.7875	3.2918

Table 7.3 *p-values for $\int_0^1 w^2(t)dt$*

Estimation of the rate of mortality

The data we have to work with is not always obtained by direct observation of cohorts. In the next lecture we will discuss somewhat "imperfect" data, that have a more complicated structure. However, in this lecture we will continue to consider a cohort, i.e. continue to consider $T_1, \ldots, T_n$ as independent and identically distributed random variables. As always, denote by F the common distribution function of each T_i. Suppose F has density f, and let μ be the rate of mortality (see (1.13)),

$$\mu(x) = \frac{f(x)}{1 - F(x)}.$$

If F is unknown, then μ is also unknown. Let us examine how we can estimate the rate of mortality $\mu(x)$ based on the observations of the cohort $T_1, \ldots, T_n$. We already know how to estimate $1 - F(x)$, and the novelty here is the problem of estimating $f(x)$. We will see that the density estimates do not converge to f as quickly as $\widehat{F}_n$ converged to F.

The most simple estimate of $f(x)$ can be built up in the following way: if $\Delta \widehat{F}_n(x) = \widehat{F}_n(x + \Delta) - \widehat{F}_n(x)$, we consider $\frac{\Delta \widehat{F}_n(x)}{\Delta}$ and then let $\Delta = \Delta_n \to 0$ for $n \to \infty$. Since the empirical distribution function $\widehat{F}_n$ does not have a density (unless we are discussing generalized functions and the like), we can not allow Δ to converge to 0 independently of n, but have to coordinate the rate of convergence of $\Delta \to 0$ with the increase of n. To choose a suitable rate for Δ let us consider the mean square deviation of the estimator

$$\widehat{f}_n(x) = \frac{\Delta \widehat{F}_n(x)}{\Delta_n} \tag{8.1}$$

from $f(x)$. To start, let us note that for each n, the expected value is

$$\mathsf{E} \frac{\Delta \widehat{F}_n(x)}{\Delta_n} = \frac{\Delta F(x)}{\Delta_n},$$

which is not $f(x)$ and becomes equal to $f(x)$ only in the limit, as $\Delta_n \to 0$. In other words, the estimate $\widehat{f}_n(x)$ is biased for each n. Then for the mean square error we have

$$\mathsf{E}\left[\widehat{f}_n(x) - f(x)\right]^2 = \mathsf{E}\left[\widehat{f}_n(x) - \frac{\Delta F(x)}{\Delta_n}\right]^2 + \left[\frac{\Delta F(x)}{\Delta_n} - f(x)\right]^2. \tag{8.2}$$

However, the variance of $\Delta \widehat{F}_n(x)$, as we know from the equation (4.8), is equal to $\Delta F(x)[1 - \Delta F(x)]/n$ and so the variance of $\widehat{f}_n(x)$ is

$$\mathsf{E}\left[\widehat{f}_n(x) - \frac{\Delta F(x)}{\Delta_n}\right]^2 = \frac{1}{n\Delta_n} \frac{\Delta F(x)}{\Delta_n}\left[1 - \Delta F(x)\right] \sim \frac{1}{n\Delta_n} f(x). \tag{8.3}$$

On the other hand

$$\frac{\Delta F(x)}{\Delta_n} - f(x) \to 0$$

no matter how $\Delta_n \to 0$. Therefore, we can draw a preliminary important conclusion:

  if $\Delta_n \to 0$, *but at the same time* $n\Delta_n \to \infty$, *then*

$$\mathsf{E}\left[\widehat{f}_n(x) - f(x)\right]^2 \to 0, \tag{8.4}$$

i.e. the estimate $\widehat{f}_n(x)$ *is consistent in the mean-square sense, and therefore, in probability:*

$$\widehat{f}_n(x) \overset{\mathsf{P}}{\to} f(x), \qquad n \to \infty.$$

If any of these conditions is not true, then there is no consistency.

Indeed, if $\Delta = \text{const}$, then the bias

$$\frac{\Delta F(x)}{\Delta} - f(x),$$

generally speaking, does not become negligible, and (8.4) is not true; and if $n\Delta_n = \text{const}$, then the variance

$$E\left[\widehat{f_n}(x) - \frac{\Delta F(x)}{\Delta_n}\right]^2 \sim \frac{1}{\text{const}} f(x)$$

does not converge to 0 and again, (8.4) is not true.

In accordance with this, and supposing that

$$\widehat{\mu}_n(x) = \frac{\widehat{f_n}(x)}{1 - \widehat{F_n}(x)},$$

we have established that

$$\widehat{\mu}_n(x) \xrightarrow{P} \mu(x),$$

if and only if $\Delta_n \to 0$ and $n\Delta_n \to \infty$.

We can clarify the question of the rate of Δ_n, if we clarify how quickly $\frac{\Delta F(x)}{\Delta}$ converges to $f(x)$ with the decrease of Δ. Suppose $F(x)$ can be differentiated in x as many times as we may need. Then, using a Taylor expansion in Δ, we obtain

$$F(x+\Delta) - F(x) = f(x)\Delta + \frac{f'(x)}{2}\Delta^2 + o(\Delta^2)$$

and

$$\frac{\Delta F(x)}{\Delta} = f(x) + \frac{f'(x)}{2}\Delta + o(\Delta).$$

So, the rate of convergence happens to equal Δ. Accept now without proof that, at the optimal choice of Δ_n, the order of magnitude of both summands in (8.2), that is, of $1/n\Delta$ and Δ^2, is the same. Equating, we obtain

$$\frac{1}{n\Delta_n} = \Delta_n^2, \quad \text{or} \quad \Delta_n = \frac{1}{n^{1/3}}, \tag{8.5}$$

which shows the order of Δ_n in n. The mean square deviation of $\widehat{f}_n(x)$ from $f(x)$ itself is of order $n^{-2/3}$:

$$E\left[\widehat{f}_n(x) - f(x)\right]^2 \sim \frac{1}{n^{2/3}}\left[f(x) + \frac{[f'(x)]^2}{4}\right]$$

differing from the order n^{-1} in the case of $E[\widehat{F}_n(x) - F(x)]^2$. As a result, we obtain

$$\widehat{\mu}_n(x) - \mu(x) = O_P(n^{-1/3})$$

again, in contrast to $\widehat{F}_n(x) - F(x) = O_P(n^{-1/2})$.

Instead of the density estimator (8.1) we can also consider a more symmetric version:

$$\widehat{\widehat{f}}_n(x) = \frac{\widehat{F}_n(x + \frac{\Delta}{2}) - \widehat{F}_n(x - \frac{\Delta}{2})}{\Delta}.$$

This estimator will have expected value

$$\frac{F(x + \frac{\Delta}{2}) - F(x - \frac{\Delta}{2})}{\Delta}$$

and variance again as in (8.3):

$$\text{Var}\,\widehat{\widehat{f}}_n(x) \sim \frac{1}{n\Delta_n}f(x).$$

However, its bias will be of smaller order:

$$F\left(x + \frac{\Delta}{2}\right) - F\left(x - \frac{\Delta}{2}\right) = f(x)\Delta + \frac{f''(x)}{3!}\cdot\frac{\Delta^3}{4} + o(\Delta^3)$$

and therefore

$$\frac{F(x + \frac{\Delta}{2}) - F(x - \frac{\Delta}{2})}{\Delta} = f(x) + f''(x)\frac{\Delta^2}{24} + o(\Delta^2),$$

so that the rate becomes Δ^2 and not Δ. Following the same principle as above, we obtain

$$\frac{1}{n\Delta_n} = \Delta_n^4, \quad \text{i.e. } \Delta_n = \frac{1}{n^{1/5}}, \tag{8.6}$$

and now Δ_n converges to 0 more slowly, while the square deviation converges to 0 more quickly than in the previous example and has the order of magnitude $n^{-4/5}$,

$$E\left[\hat{\hat{f}}_n(x) - f(x)\right]^2 \sim \frac{1}{n^{4/5}}\left[f(x) + \frac{[f''(x)]^2}{24^2}\right].$$

As a result, the estimate of the rate of mortality

$$\hat{\mu}_n(x) = \frac{\hat{\hat{f}}_n(x)}{1 - \widehat{F}_n(x)} = \frac{\widehat{F}_n(x + \frac{\Delta}{2}) - \widehat{F}_n(x - \frac{\Delta}{2})}{\Delta[1 - \widehat{F}_n(x)]}$$

converges to $\mu(x)$ somewhat more quickly:

$$\hat{\mu}_n(x) - \mu(x) = O_P(n^{-2/5}).$$

◇ **Exercise.** Verify that the choice of Δ_n, such that variance and squared bias in the right side of (8.2) are of the same magnitude, is indeed the best. What happens, for example, if, in the case of $\hat{f}_n$, we choose $\Delta_n \sim n^{-1/5}$ instead of $n^{-1/3}$? △

As we now understand, the choice of Δ reflects a certain compromise between minimizing bias and minimizing variance. One may try to choose Δ more accurately if in (8.5) and (8.6) we consider the coefficients as depending on x. We will obtain

$$\left[\frac{f'(x)}{2}\right]^2 \Delta^2 = \frac{f(x)}{n\Delta} \quad \text{and} \quad \left[\frac{f''(x)}{24}\right]^2 \Delta^4 = \frac{f(x)}{n\Delta}$$

which implies

$$\Delta_n^3 = \frac{4f(x)}{[f'(x)]^2} \cdot \frac{1}{n} \quad \text{or} \quad \Delta_n^5 = \frac{576f(x)}{[f''(x)]^2} \cdot \frac{1}{n}. \tag{8.7}$$

From these formulae we see that the optimal value of Δ_n is higher, that is, the area grouped around x is wider, the "flatter" is the density in x, i.e. the smaller is $|f'(x)|$ in the first case and $|f''(x)|$ in the second.

It is possible to continue improving the bias, i.e. making it smaller and smaller, using more and more intricate methods to extract "local information" in the neighborhood of x.

Let us present a general way of doing this. Starting with a function $K(z)$, such that

$$\int K(z)\,dz = 1,\tag{8.8}$$

transfer it into a kernel

$$\frac{1}{\Delta}K\left(\frac{x-y}{\Delta}\right).$$

This kernel is now a function of two variables, x and y, and for any x, as a function of y it is more and more "concentrated" around x as Δ becomes smaller.

Now consider the integral

$$\tilde{f}_n(x) = \frac{1}{\Delta}\int K\left(\frac{x-y}{\Delta}\right)d\widehat{F}_n(y) = \frac{1}{n\Delta}\sum_{i=1}^n K\left(\frac{x-T_i}{\Delta}\right)$$

as an estimator of a density $f(x)$ at point x. For example, if $K(z) = I_{\{-1\leq z\leq 0\}}$, then

$$\frac{1}{\Delta}K\left(\frac{x-y}{\Delta}\right) = \frac{1}{\Delta}I_{\{x\leq y\leq x+\Delta\}} \quad\text{and}\quad \tilde{f}_n(x) = \widehat{f}_n(x),$$

and if $K(z) = I_{\{-1/2\leq z\leq 1/2\}}$, then

$$\frac{1}{\Delta}K\left(\frac{x-y}{\Delta}\right) = \frac{1}{\Delta}I_{\{x-\frac{\Delta}{2}\leq y\leq x+\frac{\Delta}{2}\}} \quad\text{and}\quad \tilde{f}_n(x) = \widehat{\hat{f}}_n(x).$$

The expected value of the estimate $\tilde{f}_n(x)$ is

$$E\tilde{f}_n(x) = \frac{1}{\Delta}\int K\left(\frac{x-y}{\Delta}\right)f(y)\,dy = \int K(z)f(x-\Delta z)\,dz.$$

If $K(z)$ has a finite support, i.e. if it equals 0 outside a finite interval, then the preceding integration above is taken only over this finite interval. Therefore, we can expand $f(x-\Delta z)$ in small Δz without worrying about behavior of f on the tails. Thus we obtain

$$E\tilde{f}_n(x) = f(x)\int K(z)\,dz - f'(x)\frac{\Delta}{1!}\int K(z)z\,dz +$$

$$+ f''(x) \frac{\Delta^2}{2!} \int K(z) z^2 \, dz + \cdots$$

$$+ (-1)^m f^{(m)}(x) \frac{\Delta^m}{m!} \int K(z) z^m \, dz + o(\Delta^m).$$

Using (8.8), we see that the bias of $\widetilde{f}_n(x)$ is equal to

$$\mathsf{E}\widetilde{f}_n(x) - f(x) = -f'(x) \frac{\Delta}{1!} \int K(z) z \, dz + f''(x) \frac{\Delta^2}{2!} \int K(z) z^2 \, dz \cdots$$

$$+ (-1)^m f^{(m)}(x) \frac{\Delta^m}{m!} \int K(z) z^m \, dz + o(\Delta^m).$$

From this we understand, that if the function $K(z)$ has the additional property that

$$\int K(z) z \, dz = 0,$$

then the term with Δ disappears – and this was the case of $K(z)$ for the estimator $\widehat{f}_n$. If, moreover, $K(z)$ is such that also

$$\int K(z) z^2 \, dz = 0,$$

then the term with Δ^2 will disappear. To give an example, consider the function

$$K(z) = \frac{1}{2} \left[a I_{\{|z| \leq q\}} - b I_{\{q < |z| \leq 1\}} \right],$$

with $0 < q < 1$ and coefficients

$$a = \frac{1}{q} + \frac{q}{q+1}, \qquad b = \frac{q^2}{1 - q^2}.$$

Then

$$\int K(z) \, dz = 1, \qquad \int K(z) z \, dz = \int K(z) z^2 \, dz = 0. \qquad (8.9)$$

Therefore, the corresponding estimator $\widetilde{f}_n(x)$ has bias equal to

$$-f'''(x) \frac{\Delta^3}{3!} \int K(z) z^3 \, dz + o(\Delta^3).$$

The expression for this estimator may look somewhat unusual,

$$\widetilde{f}_n(x) = \frac{1}{2\Delta}\Big\{ a\big[\widehat{F}_n(x+q\Delta) - \widehat{F}_n(x-q\Delta)\big] - b\big[\widehat{F}_n(x+\Delta) - \widehat{F}_n(x+q\Delta)\big]$$
$$- b\big[\widehat{F}_n(x-q\Delta) - \widehat{F}_n(x-\Delta)\big]\Big\},$$

but its bias is very small.

As in (8.3), the variance of the estimator $\widetilde{f}_n(x)$ has the asymptotic form

$$\mathrm{Var}\,\widetilde{f}_n(x) = \frac{1}{n\Delta^2}\int K^2\Big(\frac{x-y}{\Delta}\Big)f(y)\,dy - \frac{1}{n\Delta^2}\bigg[\int K\Big(\frac{x-y}{\Delta}\Big)f(y)\,dy\bigg]^2$$
$$\sim \frac{1}{n\Delta}f(x)\int K^2(z)\,dz. \tag{8.10}$$

◇ **Exercise.** Ascertain that (8.10) is true, using again the change of variable

$$\frac{x-y}{\Delta} = z.$$

△

In the case of estimators $\widehat{f}_n$ and $\widehat{\widehat{f}}_n$ we had $\int K^2(z)\,dz = 1$. In the case of the kernel estimator (8.9) the integral $\int K^2(z)\,dz$ depends on the choice of q. In particular,

$$q = 0.5: \qquad a = 2.33, \qquad b = 0.33, \qquad \int K^2(z)\,dz = 1.39;$$
$$q = 0.8: \qquad a = 1.69, \qquad b = 1.78, \qquad \int K^2(z)\,dz = 1.46;$$
$$q = 0.9: \qquad a = 1.58, \qquad b = 4.26, \qquad \int K^2(z)\,dz = 2.04,$$

so that there is no advantage in the choice of q close to 1 as the variance increases.

It is quite clear now, that using the estimator $\widetilde{f}_n$, we arrive at the estimator

$$\widetilde{\mu}_n(x) = \frac{\widetilde{f}_n(x)}{1 - \widehat{F}_n(x)}$$

for the force of mortality.

The material so far can be used as an introduction to statistical smoothing methods. The need for compromise, and balance, such as that described in statements (8.6) and (8.7), forms a foundation for many smoothing techniques. For a broad discussion of the theory we refer the reader, e.g., to monographs Silverman [1986] and Green and Silverman [1994], and with more emphasis on applied and computational side, to Wand and Jones [1995] and Härdle [1991]. The possible choice is very wide.

In demographic applications it is often the case, however, that the value of Δ is given by the nature of the available data and a statistician does not have a free choice. For example, it is typical that one is given the number, or relative frequency, of those from a cohort, who died at the age of x years, without any reference to months, and still less – to days. In other words, if $[x]$ denotes the integer part of x in years (for instance, for exact age of $x = 72$ years, 8 months and 12 days, $[x] = 72$), then a statistician has to operate with values $\widehat{F}_n([x])$. Consider, therefore, the ratio

$$\frac{\Delta\widehat{F}_n(k)}{1 - \widehat{F}_n(k)}, \qquad k \text{ is an integer,} \qquad (8.11)$$

where $\Delta\widehat{F}_n(k) = \widehat{F}_n(k+1) - \widehat{F}_n(k)$. Since in this case Δ is fixed and equals 1, expression (8.11) coincides with the estimator of the rate of mortality $\widehat{\mu}_n(k)$ with $\Delta = 1$. If we need to obtain an estimation of $\mu(x)$ at some x between k and $k+1$, and, in particular, for the mid-point $x = k + 1/2 = k$ years and 6 months, for estimation of $f(x)$ there is no benefit in taking anything else but $\Delta\widehat{F}_n(k)$, as this increment has the smallest bias and is nearest to $f(x)$ in a mean square sense, see (8.2), exactly for the mid-point $x = k + 1/2$. However, instead of $\widehat{F}_n(k)$ in the denominator it is preferable to use a linear combination

$$(k+1-x)\widehat{F}_n(k) + (x-k)\widehat{F}_n(k+1)$$

as an estimator for $F(x)$. This linear combination has the smallest bias among all linear combinations and the smallest mean square error.

It will be very useful for the future to note the relationship between the estimator (8.11) of $\mu(x)$ and the empirical distribution function:

$$1 - \widehat{F}_n(k) = \prod_{j=0}^{k-1}\left[1 - \frac{\Delta\widehat{F}_n(j)}{1 - \widehat{F}_n(j)}\right]. \tag{8.12}$$

Understanding $d\widehat{F}_n(x)$ as 1, if $x = T_i$, and as 0, if $x \neq T_i$, $i = 1,2\ldots,n$, one can write a continuous analogue of (8.12):

$$1 - \widehat{F}_n(x) = \prod_{y \leq x}\left[1 - \frac{d\widehat{F}_n(y)}{1 - \widehat{F}_n(y)}\right].$$

This continuous product is what replaces the relationship

$$1 - F(x) = e^{-\int_0^x \mu(y)dy}$$

in the case when F is discrete and, therefore, has no density. We shall use a similar relationship in a more general situation in the Lecture 9.

Lecture 9

Censored observations. Related point processes

In this lecture we again consider lifetimes of individuals from the same cohort, but now our observations are more complicated and reflect realistic situations we often encounter in real life.

As we mentioned before, a cohort may consist of people of the same generation, of the same gender and of similar health. In clinical trials, for example, they often study the group of patients in more or less the same initial conditions who had undergone approximately the same treatment, say, the same type of operation.

So, there is a sequence of n independent identically distributed lifetimes $T_1, T_2, \ldots, T_n$, which we begin to observe from a certain moment t_0. This t_0 is not important for us and we can always assume that it equals 0. Otherwise, instead of $T_1, T_2, \ldots, T_n$ we can simply use the new lifetimes, or remaining lives, $T_1 - t_0, T_2 - t_0, \ldots, T_n - t_0$. Consider the situation when we are not able to observe all the lifetimes because some of the observed persons leave the field at some random moments of time. That is, assume that there is another set of independent and identically distributed random variables $Y_1, Y_2, \ldots, Y_n$, called censoring observations, such that for an each i-th person we observe not T_i but only

$$\widetilde{T}_i = \min(T_i, Y_i).$$

Therefore, what we observe is either duration of life of the i-th person, if the censoring has not yet occurred, or just a censoring variable Y_i.

Apart from $\widetilde{T}_i$ the researcher also knows the indicators

$$\delta_i = I_{\{T_i = \widetilde{T}_i\}} = \begin{cases} 1, & \text{if } T_i = \widetilde{T}_i, \text{ i.e. there was no censoring;} \\ 0, & \text{if } T_i > \widetilde{T}_i, \text{ i.e. censoring has occurred;} \end{cases}$$

which means it is known whether the observation was censored or not.

As usual, we will assume that the distribution function of each lifetime T_i has a density f, and in this case one can ignore the events $T_i = Y_i$; if the density exists, these events have probability 0.

Consider now two point processes, similar to the binomial process $z_n(t)$ we introduced in Lecture 3 – see, for example, (3.2):

$$
\begin{aligned}
N_n(t) &= \sum_{i=1}^n I_{\{\widetilde{T}_i < t\}} \delta_i, && t \geq 0, \\
Y_n(t) &= \sum_{i=1}^n I_{\{\widetilde{T}_i \geq t\}}, && t \geq 0.
\end{aligned}
\tag{9.1}
$$

All the "randomness" in these processes is contained in the random moments, the "points", at which they jump, hence the name.

The value of $N_n(t)$ shows how many uncensored lifetimes have been observed before the time t. Note that it could not have been written as

$$N_n(t) = \sum_{i=1}^n I_{\{T_i < t\}}, \quad t \geq 0,$$

because T_i we do not always observe. The value of $Y_n(t)$ shows how many individuals, out of the initial n, are still to be observed, or remain "at risk". Using processes $N_n(t)$ and $Y_n(t)$, $t \geq 0$, we want to derive an estimator for the distribution function F of the lifetimes.

If we had no censoring, then $N_n(t)$ would be just the binomial process $z_n(t)$ and $Y_n(t)$ would be equal to $n - z_n(t)$ and hence would contain nothing new relative to $z_n(t)$. Then as an estimator of $F(t)$ one would certainly use the empirical distribution function. With the censorship, $N_n(t) + Y_n(t)$ becomes random and we only know that $N_n(t) + Y_n(t) \leq n$, and so what should assume the role of the empirical distribution function is not yet clear.

At first glance, it is, perhaps, possible to say that if the observation is censored, and we know only Y_i but not T_i, then such observation could have been disregarded and thrown away. However, this will be very inefficient: even though we do not know the true value of T_i in such cases, we still know that T_i was greater than Y_i and this carries some information about the distribution of T_i, which should not be ignored.

◇ **Exercise.** Suppose we observed a sequence $\{\widetilde{T}_i, \delta_i\}$ for four persons, $i = 1, \ldots, 4$, which, in years, happened to be

$\widetilde{T}_1 = 76.2$, $\delta_1 = 1$, i.e., the first person lived for 76.2 years and there was no censorship;

$\widetilde{T}_2 = 57.3$, $\delta_2 = 0$, i.e., the second person left the observation field at the age of 57.3 years, and an observer only knows that his lifetime was longer than 57.3 years;

$\widetilde{T}_3 = 64.6$, $\delta_3 = 1$;

$\widetilde{T}_4 = 71.6$, $\delta_4 = 0$.

Draw the trajectory of $N_n(t)$ and $Y_n(t)$ for this data. △

An assumption that an observation of T_i could be censored is, certainly, quite realistic: people are moving to other places, are changing their affiliations with medical institutions, are leaving for good. Another example of censorship: suppose a person holds insurance against an event A, say, has a life insurance under some conditions, and let T be a random time until A occurs. If, however, the person died earlier at the time Y from an accident, then a statistician could observe only $Y = \widetilde{T}$. Similar situations occur every now and then in reliability problems: if a device has two components, I and II, such that failure of one of them leads to the failure of the device, and if T is the time until failure of component I and Y is the time until failure of component II, then at the failure of the device we can only observe $\min(T, Y)$, and we can know which of the components failed, that is, we can know $\delta = I_{\{T < Y\}}$.

An assumption that the censoring moments $Y_1, Y_2, \ldots, Y_n$ are random, independent of each other and independent from $T_1, T_2, \ldots, T_n$, and have the same distribution, is often also realistic. Indeed, why should a departure of one patient depend on the departure of another patient? And why should some departure times be more likely for one patient than for another? Therefore we will use this assumption below as well. Denote the distribution function of each Y_i by G:

$$P\{Y_i \le x\} = G(x).$$

◇ **Exercise.** Verify that

$$P\{\widetilde{T}_i \ge t\} = [1 - F(t)][1 - G(t)].$$

△

Consider now the processes $N_n(t)$ and $Y_n(t)$. Trajectories of the process $N_n(t)$ are non-decreasing piece-wise constant functions and can be moving upward at all t when n increases. To extract a "stable part" from $N_n(t)$ we need to center and normalize it appropriately. Since

$$E I_{\{\widetilde{T}_i < t\}} \delta_i = P\{\widetilde{T}_i < t, T_i < Y_i\} = P\{T_i < t, T_i < Y_i\} = \int_0^t [1 - G(y)] dF(y)$$

and all pairs $(\widetilde{T}_i, \delta_i)$ have the same distribution, then

$$E N_n(t) = n \int_0^t [1 - G(y)] dF(y).$$

One could develop an asymptotic theory for the processes $N_n(t)$ with this centering, but we much prefer to consider a centering of a different, more flexible sort, which we now present.

For a small increment $\Delta N_n(t) = N_n(t + \Delta) - N_n(t)$ its expected value is, certainly,

$$E \Delta N_n(t) = n \int_t^{t+\Delta} [1 - G(y)] dF(y) \sim n[1 - G(t)] \Delta F(t).$$

However, let us consider instead the conditional expected value of this increment:

$$E \left[\Delta N_n(t) \mid \text{values of all } \widetilde{T}_i < t \text{ and their corresponding } \delta_i \right].$$

Actual calculation of this expected value is easy. To do this one can consider an increment of just one summand of $N_n(t)$, see (9.1), i.e. consider

$$E\left[I_{\{t\le \widetilde{T}_i<t+\Delta\}}\delta_i \mid I_{\{\widetilde{T}_i\ge s\}} \text{ and } I_{\{\widetilde{T}_i<s\}}\delta_i, \ s\le t\right]. \tag{9.2}$$

Knowing the trajectory of $I_{\{\widetilde{T}_i<s\}}$ for $s\le t$ we either know that $\widetilde{T}_i \ge t$, or we know the value of $\widetilde{T}_i$ for $\widetilde{T}_i < t$. In the first case the trajectory of $I_{\{\widetilde{T}_i<s\}}\delta_i$, $s\le t$, does not bring in new information, but in the second case we gain the value of δ_i. It is clear that if $\widetilde{T}_i < t$, conditional expected value (9.2) equals 0, because $\widetilde{T}_i$ cannot then occur between t and $t+\Delta$, but if $\widetilde{T}_i \ge t$, then the conditional expected value (9.2) equals

$$P\{t \le \widetilde{T}_i < t+\Delta, \ \delta_i = 1 \mid \widetilde{T}_i \ge t\}$$
$$= \frac{\int_t^{t+\Delta}[1-G(y)]dF(y)}{[1-F(t)][1-G(t)]} \sim \frac{[1-G(t)]\Delta F(t)}{[1-G(t)][1-F(t)]} = \mu(t)\Delta + o(\Delta) \tag{9.3}$$

and, therefore,

$$E\left[I_{\{t\le \widetilde{T}_i<t+\Delta\}}\delta_i \mid I_{\{\widetilde{T}_i\ge s\}}, \ I_{\{\widetilde{T}_i<s\}}\delta_i, \ s\le t\right] = I_{\{\widetilde{T}_i\ge t\}}\mu(t)\Delta + o(\Delta). \tag{9.4}$$

If in condition in (9.2) one adds other events based on $\widetilde{T}_j, \delta_j$ with $j\neq i$, the expression of (9.3) will not change because $(\widetilde{T}_i,\delta_i)$ does not depend on the other pairs. Therefore, finally,

$$E[\Delta N_n(t) \mid I_{\{\widetilde{T}_i\ge s\}} \text{ and } I_{\{\widetilde{T}_i<s\}}\delta_i, \ s\le t, \ i=1,\ldots,n]$$
$$= Y_n(t)\mu(t)\Delta + o(\Delta).$$

The key fact here is that, unlike the unconditional expected value, this expression does not include the distribution function G. In our problem we want to estimate, or derive some other inference, about unknown F, while G, equally unknown, is simply a so called "nuisance parameter". Due to conditioning, it disappeared in (9.3) (see, e.g., Aalen et al. [2009]).

Speaking heuristically, the conditional expected value we derived is the "best forecast" of the value of $\Delta N_n(t)$ based on the past of the processes N_n and Y_n up to the moment t, while the difference

$$\Delta N_n(t) - Y_n(t)\mu(t)\Delta \tag{9.5}$$

is a "purely random", "unpredictable" part of $\Delta N_n(t)$.

Rewrite the process in integral form, that is, integrate the difference above from 0 to the current t:

$$M_n(t) = N_n(t) - \int_0^t Y_n(y)\mu(y)\,dy. \tag{9.6}$$

The conditional expected value of the difference (9.5) is, as we have seen, $o(\Delta)$. This property is completely retained in the limit: not only is $\mathsf{E}M_n(t) = 0$ for all t, but also
  for all $t' > t \geq 0$

$$\mathsf{E}[M_n(t') - M_n(t) \mid Y_n(s), N_n(s), s \leq t] = 0. \tag{9.7}$$

This is an important property, and shows that the process $N_n(t)$, when centered as in (9.5), becomes a martingale (see the upcoming definition). The centering process

$$\int_0^t Y_n(y)\mu(y)\,dy$$

is called the compensator of N_n.

We prove (9.7) in two steps: first we recall, that, as we essentially have seen in (9.3),

$$\mathsf{E}[N_n(t') - N_n(t) \mid Y_n(s), N_n(s), s \leq t]$$
$$= Y_n(t)\frac{\int_t^{t'}[1 - G(y)]\,dF(y)}{[1 - F(t)][1 - G(t)]}, \tag{9.8}$$

and then we prepare the expression for

$$\mathsf{E}[Y_n(y) \mid Y_n(s), N_n(s), s \leq t].$$

If this turns out to be the right-hand side of (9.8), the martingale property will be proved.
  Note again, that

$$\mathsf{E}\left[I_{\{t \leq \tilde{T}_i < y\}} \mid I_{\{\tilde{T}_i \geq s\}}, I_{\{\tilde{T}_i < s\}}\delta_i, s \leq t\right]$$

$$= \mathsf{P}\left\{I_{\{t \le \tilde{T}_i < y\}} \mid I_{\{\tilde{T}_i \ge t\}}\right\} = I_{\{\tilde{T}_i \ge t\}} \frac{H(y) - H(t)}{1 - H(t)}$$

where $H(y)$ denotes the distribution function of $\tilde{T}_i$, i.e.

$$1 - H(y) = [1 - F(y)][1 - G(y)],$$

so that

$$\mathsf{E}[Y_n(y) \mid Y_n(s), N_n(s), s \le t] = -Y_n(t) \frac{H(y) - H(t)}{1 - H(t)} + Y_n(t)$$

$$= Y_n(t) \frac{1 - H(y)}{1 - H(t)}. \qquad (9.9)$$

From the last equality it follows that

$$\mathsf{E}\left[\int_t^{t'} Y_n(y)\mu(y)\,dy \mid Y_n(s), N_n(s), s \le t\right]$$

$$= \int_t^{t'} \mathsf{E}[Y_n(y) \mid Y_n(s), N_n(s), s \le t]\mu(y)\,dy$$

$$= Y_n(t) \int_t^{t'} \frac{[1 - F(y)][1 - G(y)]}{[1 - F(t)][1 - G(t)]} \frac{f(y)}{1 - F(y)}\,dy$$

$$= Y_n(t) \frac{\int_t^{t'}[1 - G(y)]\,dF(y)}{[1 - F(t)][1 - G(t)]},$$

which coincides with the right-hand side of (9.8). Therefore

$$\mathsf{E}[M_n(t') - M_n(t) \mid Y_n(s), N_n(s), s \le t]$$

$$= \mathsf{E}[N_n(t') - N_n(t) \mid Y_n(s), N_n(s), s \le t]$$

$$- \mathsf{E}\left[\int_t^{t'} Y_n(y)\mu(y)\,dy \mid Y_n(s), N_n(s), s \le t\right] = 0,$$

and hence (9.7) is proved.

To better align what we have obtained so far with the theory of martingales (see, e.g., Klebaner [2005]) we need more general terminology. Denote by $\mathscr{F}_t^n$ the σ-algebra of events, generated by the processes $Y_n(s)$ and $N_n(s)$ for $s \le t$,

$$\mathscr{F}_t^n = \sigma\{Y_n(s), N_n(s), s \le t\}. \qquad (9.10)$$

That is, any event[1], which can be expressed in terms of $Y_n(s)$ and $N_n(s)$ up to the moment t, is included in $\mathscr{F}_t^n$. One can think of $\mathscr{F}_t^n$ as the "history" of the processes Y_n and N_n up to the moment t. An elementary event, or a point, in σ-algebra $\mathscr{F}_t^n$ is the pair of trajectories of Y_n, N_n for $s \leq t$. It is clear that

$$\mathscr{F}_t^n \subset \mathscr{F}_{t'}^n, \quad t \leq t',$$

i.e. the family of σ-algebras $\{\mathscr{F}_t^n\}_{t \geq 0}$ is increasing in t (with respect to inclusion). Any increasing family of σ-algebras $\{\mathscr{F}_t\}_{t \geq 0}$ is called a filtration. A process $M(t)$, $t \geq 0$, is called a martingale with respect to the given filtration $\{\mathscr{F}_t\}_{t \geq 0}$, if

$$\mathsf{E}|M(t)| < \infty \quad \text{for } t < \infty$$

and for all $t' > t$

$$\mathsf{E}[M(t') - M(t) | \mathscr{F}_t] = 0.$$

The proof of the first condition for our process M_n we leave as an exercise, while the second condition is just what we proved in (9.7). For convenience, we reformulate it as follows:

the process M_n, defined by (9.6), is a martingale with respect to the filtration $\{\mathscr{F}_t^n\}_{t \geq 0}$, defined in (9.10), i.e.

$$\mathsf{E}|M_n(t)| < \infty \quad \text{for } t < \infty$$

(see the following exercise) *and*

$$\mathsf{E}[M_n(t') - M_n(t) | \mathscr{F}_t^n] = 0$$

(see (9.7)).

◇ **Exercise.** a) What is the form of the martingale M_n, if $n = 1$?

b) What is the form of the martingale M_n, if there is no censoring?

---

[1] That is, all "good" events, such that we can define their probability in a unique way; there are events, called "non-measurable", for which we cannot do this, but we omit discussion of this here.

c) It is clear that $N_n(t) + Y_n(t) \le n$. What happens in the case b)?

d) As a result, find an expression for M_n for the binomial point process z_n.

e) Is it true that $E|M_n(t)| < \infty$ for all t? $\triangle$

◇ **Exercise.** We have seen that

$$EN_n(t) = n \int_0^t [1 - G(y)] dF(y),$$

and it is easy to show (see the second exercise in this lecture), that

$$EY_n(t) = n[1 - G(t)][1 - F(t)].$$

Suppose we center $N_n(t)$ and $Y_n(t)$ by their expected values above and substitute the centered processes

$$N_n(t) - n \int_0^t [1 - G(y)] dF(y) \quad \text{and} \quad Y_n(t) - n[1 - G(t)][1 - F(t)]$$

in M_n instead of $N_n(t)$ and $Y_n(t)$, respectively. What will happen to the form of M_n? $\triangle$

Lecture 10

Kaplan–Meier estimator (product-limit estimator) for F

Now we can introduce the desired estimator for the distribution function F of lifetimes, based on censored observations.

Since M_n accumulates the "purely noisy" component of N_n, we obtain our estimator by equating M_n to zero, that is, as a solution to the equation

$$N_n(t) - \int_0^t Y_n(y) \frac{dF(y)}{1 - F(y)} = 0, \qquad 0 \le t \le \widetilde{T}_{(n)}, \qquad (10.1)$$

with respect to F. Here $\widetilde{T}_{(n)}$ is the moment of the last jump of the process Y_n.

For piece-wise constant function $N_n(t)$ with jumps, and all trajectories of $N_n(t)$ are such functions, the solution is given by

$$1 - \overline{F}_n(t) = \prod_{y < t} \left(1 - \frac{dN_n(y)}{Y_n(y)}\right), \qquad 0 \le t \le \widetilde{T}_{(n)}.$$

The function $\overline{F}_n$ is called the Kaplan–Meier estimator, or product-limit estimator, for the distribution function F. It is clear that $dN_n(y) = 1$, if $y = T_i$ and $\delta_i = 1$ for some $i = 1, \ldots, n$, and $dN_n(y) = 0$ for all other values of y. At the last point $y = \widetilde{T}_{(n)}$ we have $Y_n(\widetilde{T}_{(n)}) = 1$. It is also clear that $\overline{F}_n$ itself is a piece-wise constant function, which jumps at the same points as N_n does, and that these jumps are equal

$$d\overline{F}_n(t) = \lim_{\Delta t \to 0} \left(\overline{F}_n(t + \Delta t) - \overline{F}_n(t)\right)$$

$$= \prod_{y<t}\left(1-\frac{dN_n(y)}{Y_n(y)}\right)\lim_{\Delta t\to 0}\left[1-\prod_{t\le y<t+\Delta t}\left(1-\frac{dN_n(y)}{Y_n(y)}\right)\right]$$

$$= \left[1-\overline{F}_n(t)\right]\frac{dN_n(t)}{Y_n(t)}.$$

Hence the integral in (10.1) taken with respect to $\overline{F}_n$ is indeed equal to $N_n(t)$.

◇ **Exercise.** Verify that if there is no censoring, then $\overline{F}_n(s)$ becomes the empirical distribution function $\widehat{F}_n$ based on $T_1,\dots,T_n$. △

◇ **Exercise.** It could be tempting to say that from (10.1) there follows

$$\frac{dN_n(t)}{Y_n(t)}-\frac{dF(t)}{1-F(t)}=0$$

or

$$\int_0^t\frac{dN_n(y)}{Y_n(y)}+\ln[1-F(t)]=0.$$

From the latter there would have followed a different expression

$$1-\overline{F}_n(t)=e^{-\int_0^t dN_n(y)/Y_n(y)}. \tag{10.2}$$

To see why this is not correct verify the following:

a) If F is differentiable, then (10.1) cannot be true for all t.

b) If F is not differentiable, then

$$\int_0^t\frac{dF(y)}{1-F(y)}\neq-\ln[1-F(t)].$$

△

◇ **Exercise.** Continuous product is an interesting object, but in the case of $\overline{F}$ we could do without it, because

$$\prod_{y<t}\left(1-\frac{dN_n(y)}{Y_n(y)}\right)=\exp\int_0^t\ln(1-\frac{1}{Y_n(y)})dN_n(y),\quad 0\le t\le \tilde{T}_{(n)}.$$

and one could use the integral in the exponent to describe $\overline{F}$.
Verify this equality and compare with (10.2) to see again where the "rub" is: in $1/Y_n(y)$ not being small to justify expansion of the $\ln(1-1/Y_n(y))$ up to linear term only.

◇ **Exercise.** Suppose in an experiment ten values of $\tilde{T}_i$ and corresponding δ_i were obtained:

$$(72.6, 1), \quad (64.3, 1), \quad (59.3, 0), \quad (74.8, 0), \quad (42.7, 1),$$
$$(79.1, 1), \quad (68.2, 1), \quad (71.5, 0), \quad (57.7, 0), \quad (66.3, 1),$$

that is, four observation out of these ten have been censored.

a) Draw the trajectory of $\overline{F}_n$ based on this data. Is $\overline{F}_n$ a proper distribution function, that is, does $F_n(t)$ reach 1 for $t \geq \tilde{T}_{(n)}$?

b) Change the values of δ_i, $i = 1, \ldots, 10$, so that $\overline{F}_n(t)$ remains < 1 for all t. Formulate a necessary and sufficient condition for $\overline{F}_n(t)$ to reach 1 for sufficiently large t. △

To be able to judge how accurate the estimator $\overline{F}_n$ is, we need to know the behavior of the difference

$$\overline{F}_n(t) - F(t), \qquad t \geq 0,$$

or a normalized version of it, as a random process when n is large. It would be useful to know this behavior also in the problem of testing the hypothesis

$$F = F_0$$

for a given hypothetical F_0 (replacing F with F_0). However, in this latter case, instead of the difference $\overline{F}_n - F_0$ we can use the martingale M_n itself (with F replaced by F_0), and study the asymptotic behavior of M_n for large n. This is easily furnished by the central limit theorem for martingales (see Rebolledo [1980], also Shorack and Wellner [2009], ch.26).

Namely, introduce the so-called quadratic variation process $\langle M_n \rangle(t)$ of the martingale $M_n(t)$. According to its descriptive definition $\langle M_n \rangle$ is such a process that the difference

$$M_n^2(t) - \langle M_n \rangle(t)$$

is again a martingale. It is not difficult, however, to construct it. For this consider

$$\mathsf{E}\left[(\Delta M_n(t))^2 \mid \mathscr{F}_t^n \right]$$

and extract the main term in Δ of this conditional variance and then integrate it. This integral will be $\langle M_n \rangle (t)$.

Using the independence of $(\widetilde{T}_i, \delta_i)$, $i = 1, \dots, n$, we obtain

$$\mathsf{E}\left[(\Delta M_n(t))^2 \mid \mathscr{F}_t^n \right] = \sum_{i=1}^{n} \mathrm{Var}\left[I_{\{t \leq \widetilde{T}_i < t+\Delta\}} \delta_i \mid I_{\{\widetilde{T}_i \geq s\}}, I_{\{\widetilde{T}_i < s\}} \delta_i, \, s \leq t \right]$$

$$= \sum_{i=1}^{n} \left\{ \mathsf{E}\left[I_{\{t \leq \widetilde{T}_i < t+\Delta\}}^2 \delta_i^2 \mid I_{\{\widetilde{T}_i \geq s\}}, I_{\{\widetilde{T}_i < s\}} \delta_i, \, s \leq t \right] \right.$$

$$\left. - \left(\mathsf{E}\left[I_{\{t \leq \widetilde{T}_i < t+\Delta\}} \delta_i \mid I_{\{\widetilde{T}_i \geq s\}}, I_{\{\widetilde{T}_i < s\}} \delta_i, \, s \leq t \right] \right)^2 \right\}.$$

Now making use of the fact that

$$I_{\{t \leq \widetilde{T}_i < t+\Delta\}}^2 \delta_i^2 = I_{\{t \leq \widetilde{T}_i < t+\Delta\}} \delta_i,$$

and of the asymptotic expression (9.4) we find that each summand in the right hand side equals

$$\mathsf{E}\left[I_{\{t \leq \widetilde{T}_i < t+\Delta\}} \delta_i \mid I_{\{\widetilde{T}_i > s\}}, I_{\{\widetilde{T}_i \leq s\}} \delta_i \right]$$

$$- I_{\{\widetilde{T}_i \geq t\}} \mu^2(t) \Delta^2 + o(\Delta^2) = I_{\{\widetilde{T}_i \geq s\}} \mu(t) \Delta + o(\Delta).$$

Therefore

$$\mathsf{E}\left[(\Delta M_n(t))^2 \mid \mathscr{F}_t^n \right] = Y_n(t) \mu(t) \Delta + o(\Delta).$$

Integrating the main term we obtain

$$\langle M_n \rangle(t) = \int_0^t Y_n(y) \mu(y) \, dy, \tag{10.3}$$

which is the process already familiar to us — the compensator of $N_n(t)$. The fact, that the compensator of a point process N_n is equal to its quadratic variation, is a general fact, true for any point process (see, e.g., Karr [1991], Bremaud [1981]).

◇   **Exercise.** A binomial random variable ν_n with number of trials Y_n and probability of success p, $\nu_n \sim \mathrm{bin}(\cdot, Y_n, p)$, has expected value $\mathsf{E}\nu_n = Y_n p$ and variance $\mathrm{Var}\,\nu_n = Y_n p(1-p)$. These quantities have the same order of magnitude when $p \to 0$. Is this analogous to the similarity between the compensator and $\langle M_n \rangle$?   △

◇ **Exercise.** Let $\xi(t)$, $t \geq 0$, denote a time homogeneous Poisson process with intensity λ. Is the process $L(t) = \xi(t) - \lambda t$ a martingale? If so, derive an expression for $\langle L \rangle(t)$. △

Let us now formulate a central limit theorem for martingales, without proof, then verify its conditions for our martingales M_n, normalized by $1/\sqrt{n}$. Namely (see Rebolledo [1980], Shorack and Wellner [2009]),

let $m_n(t)$, $t \geq 0$, $n = 1, 2 \ldots$ be a sequence of martingales, such that
(a) (Lindeberg condition)

$$\sum_{0 \leq t < \infty} E[\delta m_n(t)]^2 I_{\{|\delta m_n(t)|^2 > \varepsilon\}} \xrightarrow{P} 0, \qquad n \to \infty,$$

where $\delta m_n(t) = m_n(t+) - m_n(t-)$ denotes the jump of m_n at t,
and
(b)

$$\langle m_n \rangle(t) \xrightarrow{P} A(t), \qquad n \to \infty,$$

where $A(t)$, $t \geq 0$, is a deterministic non-decreasing function. Then

$$m_n \xrightarrow{d} W,$$

where W is Brownian motion in time A (cf. (5.1) and (5.2) in Lecture 5).

From this theorem immediately follows the limit theorem for $\frac{1}{\sqrt{n}} M_n$:

for martingales M_n, defined by (9.6), the following convergence is true:

$$\frac{1}{\sqrt{n}} M_n \xrightarrow{d} W, \tag{10.4}$$

where

$$A(t) = \int_0^t [1 - G(y)] \, dF(y) \tag{10.5}$$

and W is Brownian motion in time A.

Indeed, the Lindeberg condition can be verified easily: since $0 \leq \delta N_n(t) \leq 1$, we have that

$$\delta m_n(t) = \frac{1}{\sqrt{n}} \delta N_n(t) \leq \frac{1}{\sqrt{n}},$$

so that $I_{\{|\delta m_n(t)|^2 > \varepsilon\}} = 0$ for all t, as soon as $1/n < \varepsilon$, and the sum becomes equal to 0. To establish (b), note that

$$\left\langle \frac{1}{\sqrt{n}} M_n \right\rangle(t) = \frac{1}{n}\langle M_n \rangle(t) = \int_0^t \frac{Y_n(s)}{n} \mu(s)\, ds. \qquad (10.6)$$

But

$$\frac{Y_n(s)}{n} = \frac{1}{n}\sum_{i=1}^n I_{\{\widetilde{T}_i \geq s\}}$$

is just the tail of the empirical distribution function based on independent and identically distributed random variables $\widetilde{T}_i$. Therefore from the Glivenko–Cantelli theorem (see Lecture 3), it follows that

$$\frac{Y_n(s)}{n} \to [1 - F(s)][1 - G(s)] \qquad n \to \infty$$

uniformly in s with probability 1. Therefore, for all $t > 0$,

$$\left\langle \frac{1}{\sqrt{n}} M_n \right\rangle(t) \to \int_0^t [1 - F(s)][1 - G(s)]\, \mu(s)\, ds$$

$$= \int_0^t [1 - G(s)]\, dF(s) = A(t)$$

with probability 1, and not only in probability. Hence the limit theorem (10.4) is proved.

Let us go back to the difference $\overline{F}_n - F$. The Glivenko–Cantelli theorem for $\overline{F}_n$, that is, the statement that
for $\tau = \inf\{t : [1 - G(t)][1 - F(t)] = 0\}$,

$$\sup_{0 < t \leq \tau} |\overline{F}_n(t) - F(t)| \to 0, \qquad n \to \infty, \qquad (10.7)$$

with probability 1,
requires now some more effort than for the case of the empirical distribution function. It was proved in its final form not so long ago in Stute and Wang [1993]. If we accept it without proof, then we are otherwise ready to derive the central limit theorem for $\overline{F}_n$. In the presentation below we follow Gill [1980] and Andersen et al. [1993]. Consider the normalized ratio

$$\sqrt{n}\left[\frac{1 - \overline{F}_n(t)}{1 - F(t)} - 1 \right]. \qquad (10.8)$$

Since according to the definition of $\overline{F}_n(t)$

$$\Delta[1 - \overline{F}_n(t)] = [1 - \overline{F}_n(t)]\left[\prod_{t < y \leq t + \Delta}\left(1 - \frac{dN_n(y)}{Y_n(y)}\right) - 1\right],$$

for the increment of the process (10.8) with small Δ we obtain

$$\sqrt{n}\frac{1 - \overline{F}_n(t)}{1 - F(t)}\left[-\frac{\Delta N_n(t)}{Y_n(t)} + \frac{\Delta F(t)}{1 - F(t)}\right] + o(\Delta),$$

which after simple rearrangements leads to the following identity

$$\sqrt{n}\left[\frac{1 - \overline{F}_n(t)}{1 - F(t)} - 1\right] = -\int_0^t \frac{1 - \overline{F}_n(s)}{1 - F(s)}\frac{n}{Y_n(s)}\frac{dM_n(s)}{\sqrt{n}}. \qquad (10.9)$$

Using the Glivenko–Cantelli theorem (10.7), for all $s < \tau$ we can replace

$$[1 - \overline{F}_n(s)]/[1 - F(s)]$$

by 1 and replace also $Y_n(s)/n$ by $[1 - F(s)][1 - G(s)]$. What we will obtain is

$$\frac{\sqrt{n}[\overline{F}_n(t) - F(t)]}{1 - F(t)} \sim \int_0^t \frac{n}{Y_n(s)}\frac{dM_n(s)}{\sqrt{n}}$$

$$\xrightarrow{d} \int_0^t \frac{1}{[1 - F(s)][1 - G(s)]}dW(s), \quad t \leq \tau', \qquad (10.10)$$

and

$$\sqrt{n}[\overline{F}_n(t) - F(t)] \xrightarrow{d} [1 - F(t)]$$

$$\times \int_0^t \frac{1}{[1 - F(s)][1 - G(s)]}dW(s), \quad t \leq \tau', \qquad (10.11)$$

where the processes are considered on the interval $[0, \tau']$ with $\tau' < \tau$, and not on the whole $[0, \tau]$, and where W is the Brownian motion in time A, see (10.4) and (10.5). This last statement is the central limit theorem for the Kaplan–Meier estimator $\overline{F}_n$ that we wanted. Note that an essential ingredient here was the martingale central limit theorem for M_n.

In connection with this theorem few remarks seem to be useful. First, the integral on the right-hand side of (10.10) and (10.11) is also a Brownian motion with respect to the time

$$C(t) = \int_0^t \frac{1}{[1-F(s)]^2[1-G(s)]^2} A(ds)$$
$$= \int_0^t \frac{1}{[1-F(s)]^2[1-G(s)]} dF(s),$$

and we see, that "typically" $C(\tau) = \infty$. This partly explains why we chose to narrow the time interval in (10.10). Second, the jumps in the integral in (10.9) are of the order

$$\left| \frac{1-\bar{F}_n(s)}{1-F(s)} \right| \frac{n}{Y_n(s)} \frac{1}{\sqrt{n}},$$

which can be very large for s close to τ and problems with the Lindeberg condition will arise. This again shows why one would prefer to restrict the limit theorem to a narrower time interval. The third remark is that in the limit theorem (10.11) the unknown distribution function G is involved, so that, formally speaking, the limit distribution of the limiting process remains unknown. We discuss how to resolve this difficulty in the next lecture.

It is interesting to consider a somewhat different path to derive a convenient asymptotic expression for the difference $\sqrt{n}(\bar{F}_n - F)$. Indeed, why should we not consider this difference straightaway instead of starting with the ratio (10.8)? We have two identities

$$1 - \bar{F}_n(t) = 1 - \int_0^t [1 - \bar{F}_n(s)] \frac{dN_n(s)}{Y_n(s)}$$

and

$$1 - F(t) = 1 - \int_0^t [1 - F(s)] \mu(s) \, ds.$$

Subtracting the former from the latter and multiplying the difference by $\sqrt{n}$, one finds that

$$\sqrt{n}[\bar{F}_n(t) - F(t)] = \sqrt{n} \int_0^t \left([1 - \bar{F}_n(s)] \frac{dN_n(s)}{Y_n(s)} - [1 - F(s)] \mu(s) \, ds \right).$$

Why could not one now take $1 - \overline{F}_n$ outside the brackets, thus obtaining

$$\sqrt{n} \int_0^t [1 - \overline{F}_n(s)] \left(\frac{dN_n(s)}{Y_n(s)} - \frac{1 - F(s)}{1 - \overline{F}_n(s)} \mu(s) \, ds \right)$$

and replace the resulting ratio $(1 - F(s))/(1 - \overline{F}_n(s))$ by 1? Continuing by subtracting and adding $\mu(s)$ inside the brackets, we see that this would be incorrect, as it will lead to

$$\int_0^t [1 - \overline{F}_n(s)] \sqrt{n} \left[\frac{dN_n(s)}{Y_n(s)} - \mu(s) \, ds \right] - \int_0^t \sqrt{n} [\overline{F}_n(s) - F(s)] \mu(s) \, ds$$

and $\sqrt{n} [\overline{F}_n(s) - F(s)]$ is not small, although the difference $[\overline{F}_n(s) - F(s)]$ itself is small. Altogether, the process $\sqrt{n} [\overline{F}_n - F]$ asymptotically satisfies the equation

$$\xi(t) = \int_0^t [1 - F(s)] \frac{n}{Y_n(s)} \, dm_n(s) - \int_0^t \xi(s) \mu(s) \, ds,$$

and the solution of this equation is

$$\xi(t) = [1 - F(t)] \int_0^t \frac{n}{Y_n(s)} \, dm_n(s),$$

which again leads to (10.11).

◇ **Exercise.** Suppose at point τ the distribution function G jumps from some positive value to 1: $G(\tau) < 1$ and $G(\tau + \varepsilon) = 1$ for any $\varepsilon > 0$. Suppose also that $F(\tau + \varepsilon) < 1$ for some $\varepsilon > 0$. That is, suppose τ is the largest possible value for the censoring variables Y_i, which they can take with positive probability $1 - G(\tau)$, although lifetimes T_i can take values larger than τ. Is then $C(\tau) = \infty$? △

10.1* A note on Wiener stochastic integral

Our purpose here is to explain why the integral in (10.10) is again a Brownian motion. To do this consider a Brownian motion W given on an interval $[0, \tau']$ and suppose A is its time, that is, $EW^2(t) = A(t)$ for all

$t \in [0, \tau']$. Neither A nor τ' have to be what they have been in the main text of this lecture: just some finite point τ' and some non-decreasing function $A(t)$ on $[0, \tau']$. Suppose also that g is a deterministic function on $[0, \tau']$. We want to define the process

$$\int_0^t g(s)dW(s), \quad t \in [0, \tau'],$$

and show that it again is a Brownian motion in time

$$C(t) = \int_0^t g^2(s)dA(s).$$

In other words, we need to define this integral in such a way that for any k and any collection of k points $0 = t_0 < t_1 < \cdots < t_k < t_{k+1} = \tau'$ the integrals

$$\int_{t_j}^{t_{j+1}} g(s)dW(s), \quad j = 0, 1, \ldots, k$$

are independent zero-mean Gaussian random variables and the variance of each is

$$\text{Var} \int_{t_j}^{t_{j+1}} g(s)dW(s) = \int_{t_j}^{t_{j+1}} g^2(s)dA(s).$$

To do this we need the following assumption on the function g: there exists a sequence of piece-wise constant functions (step-functions) g_n, with finite number of steps, such that

$$\int_0^{\tau'} [g(s) - g_n(s)]^2 dA(s) \to 0, \quad n \to \infty. \tag{10.12}$$

Now suppose g_n is such a step-function, which can jump only at points

$$0 = s_{0n} < s_{1n} < \cdots < s_{nn} = \tau'.$$

For any two moments $0 \le t_1 < t_2 \le \tau'$ there exist two intervals $[s_{i_1 n}, s_{i_1+1,n}]$ and $[s_{i_2 n}, s_{i_2+1,n}]$, which contain the points t_1 and t_2, respectively; these may be the same interval if t_1 and t_2 are close to each other, but for all sufficiently large n and any fixed $t_1 < t_2$ these intervals will be different. Define the integral $\int_{t_1}^{t_2} g_n(s)dW(s)$ as the sum

$$\int_{t_1}^{t_2} g_n(s)dW(s) = g_n(s_{i_1,n})[W(s_{i_1+1,n}) - W(t_1)]$$

$$+ \sum_{s_{i_1,n} < s_{jn} < s_{i_2,n}} g_n(s_{jn}) \Delta W(s_{jn}) + g_n(s_{i_2,n})[W(t_2) - W(s_{i_2,n})],$$

where $\Delta W(s_{in}) = W(s_{i+1,n}) - W(s_{in})$. From the properties of the increments of the Brownian motion $\Delta W(s_{in})$ it follows that our integral is a zero-mean Gaussian random variable and its variance is

$$\text{Var}\left(\int_{t_1}^{t_2} g_n(s)dW(s)\right) = g_n^2(s_{i_1,n})[A(s_{i_1+1,n}) - A(t_1)]$$

$$+ \sum_{s_{i_1,n} < s_{jn} < s_{i_2,n}} g_n^2(s_{jn}) \Delta A(s_{jn}) + g_n^2(s_{i_2,n})[A(t_2) - A(s_{i_2,n})]$$

$$= \int_{t_1}^{t_2} g_n^2(s)dA(s).$$

It is also obvious that, say, two such integrals, $\int_{t_1}^{t_2} g_n(s)dW(s)$ and $\int_{t_3}^{t_4} g_n(s)dW(s)$, $t_2 \leq t_3$, are independent random variables because they involve sets of increments which are independent from each other. This is certainly true for any collection of such integrals on disjoint time intervals. Therefore, the process

$$\int_0^t g_n(s)dW(s)$$

is a Brownian motion in time

$$\int_0^t g_n^2(s)dA(s). \tag{10.13}$$

It will also be useful to note that for any two step-functions g_{1n} and g_{2n}, piece-wise constant between the points s_{jn} of the same grid, the following two properties are true: (1) – linearity in g_n, that is,

$$\int_0^t \left[\alpha_1 g_{1n}(s) + \alpha_2 g_{2n}(s)\right]dW(s)$$

$$= \alpha_1 \int_0^t g_{1n}(s)dW(s) + \alpha_2 \int_0^t g_{2n}(s)dW(s)$$

for constant α_1 and α_2, which follows simply from the linearity of sum; and (2) – that (10.13) can be easily generalized to

$$\text{E} \int_0^t g_{1n}(s)dW(s) \int_0^t g_{2n}(s)dW(s) = \int_0^t g_{1n}(s)g_{2n}(s)dA(s). \tag{10.14}$$

Now note that without loss of generality, we can assume that if $n < m$ then the collection of jump points of g_m includes the jump points of g_n: indeed, given a collection of jump points of g_m, we can always add to it all jump points of g_n and in the resulting collection we can allow g_n (and g_m) to take equal values, if necessary, at neighboring points. In this way at some jump points the function g_n (and g_m) will have no jump, but this does not matter, while we can now say that g_n and g_m are piece-wise constant in between the same grid points. It follows that the integrals $\int_0^t [g_n(s) - g_m(s)] dW(s)$ are defined for all $t \leq \tau'$ and that

$$\int_0^t \left[g_n(s) - g_m(s) \right] dW(s) = \int_0^t g_n(s) dW(s) - \int_0^t g_m(s) dW(s).$$

Moreover, from (10.13) it follows that

$$\operatorname{Var}\left(\int_0^t \left[g_n(s) - g_m(s) \right] dW(s) \right) = \int_0^t \left[g_n(s) - g_m(s) \right]^2 dA(s),$$

and (10.12) implies that the integral on the right-hand side converges to 0 as soon as $n, m \to \infty$. Therefore, the variance of the difference

$$\operatorname{Var}\left[\int_0^t g_n(s) dW(s) - \int_0^t g_m(s) dW(s) \right] \to 0, \text{ as } n, m \to \infty.$$

We take without proof now that if a sequence of random variables Z_n, $n = 1, 2, \ldots$, is such that $\operatorname{Var}[Z_n - Z_m] \to 0$, as $n, m \to \infty$, then there exists a random variable Z, such that $\operatorname{Var}[Z_n - Z] \to 0$ as $n \to \infty$; and if all but finitely many Z_n are Gaussian, then Z is also Gaussian. Applying this to the integrals $\int_0^t g_n(s) dW(s)$, we conclude that there is a limiting random variable, which we denote $\int_0^t g(s) dW(s)$, such that

$$\operatorname{Var}\left[\int_0^t g_n(s) dW(s) - \int_0^t g(s) dW(s) \right] \to 0, \quad n \to \infty.$$

It is obvious that

$$\int_0^t g(s) dW(s)$$

inherits the properties of $\int_0^t g_n(s) dW(s)$, and, in particular, that

$$\int_0^t \left[\alpha_1 g_1(s) + \alpha_2 g_2(s) \right] dW(s)$$

$$= \alpha_1 \int_0^t g_1(s)dW(s) + \alpha_2 \int_0^t g_2(s)dW(s)$$

and that

$$\mathrm{Var}\left[\int_0^t g(s)dW(s)\right] = \int_0^t g^2(s)dA(s),$$

as well as

$$\mathsf{E}\int_0^t g_1(s)dW(s)\int_0^t g_2(s)dW(s) = \int_0^t g_1(s)g_2(s)dA(s).$$

◊ **Exercise.** a) The last displayed formula is important in several ways. To get better used to it, derive (10.14) from the definition of the integral $\int_0^t g_n(s)dW(s)$ and then view the last display as its limit; alternatively, derive this equality by considering

$$\mathrm{Var}\left[\int_0^t (g_1(s) + g_2(s))dW(s)\right]$$

b) Prove that the process $\int_0^t g(s)dW(s)$ has independent increments by considering the covariance between

$$\int_0^{\tau'} g_1(s)dW(s) \quad \text{and} \quad \int_0^{\tau'} g_2(s)dW(s)$$

and using

$$g_1(s) = g(s)I_{\{s<t_1\}} \quad g_2(s) = g(s)I_{\{t_1 \leq s<t_2\}}.$$

△

Statistical inference about F, based on the Kaplan–Meier estimator

Before we consider three specific statistical problems let us quickly look back on what was done in Lectures 9 and 10. The key model derived there was that if we have censored observations, and if Y_n is the process of those "at risk", then the process M_n defined as

$$M_n(t) = N_n(t) - \int_0^t Y_n(s)\mu(s)ds, \quad 0 \le t < \tau,$$

is a martingale with respect to the filtration $\{\mathscr{F}_t^n, 0 \le t < \tau\}$, where each σ-algebra is generated by the past of Y_n and N_n up to the moment t:

$$\mathscr{F}_t^n = \sigma\{Y_n(s), N_n(s), s \le t\}. \tag{11.1}$$

This was done in the context where $N_n(t)$ counted the number of lifetimes, or failure times, up to the moment t, which happened to be uncensored, while $Y_n(t)$ counted the number of those in the initial population of n, who are still under observation and "at risk" of failing.

Although in the case of censored observations we could state many facts on the asymptotic behavior of Y_n, we actually used only the one very simple property that

$$\frac{Y_n(t)}{n} \to y(t), \quad n \to \infty, \tag{11.2}$$

uniformly in t on our interval $[0, \tau]$, and the key observation about y

was that although this limiting function exists, we do not know it and do not want to formulate hypotheses about it or otherwise be obliged to specify it.

However, the same situation can exist in a much more general context. Indeed, for each moment t we may have a group of $Y_n(t)$ individuals, and each of them may fail in a small time interval $[t, t + dt)$ with the same probability $\mu(t)dt + o(dt)$, so that all individuals in our group have the same failure rate $\mu(t)$. This is an assumption, but it may be a reasonable assumption in many situations. As a consequence, the process M_n above is still a martingale with respect to the filtration defined by (11.1).

We do not need here to assume that $Y_n(t)$ is decreasing. For example $Y_n(t)$ may be a number of specific patients with a particular condition, who did not fail up to the moment t, and as new patients may arrive, $Y_n(t)$ does not have to be a decreasing process. As a matter of fact, we do not need to require any specific properties from $Y_n(t)$. Indeed, looking at the proof of the central limit theorem for M_n we can easily see that the only property we needed was the law of large number-type statement (11.2); if we assume that it is true and the integral

$$A(t) = \int_0^t y(s)\mu(s)ds$$

is finite for all $t \leq \tau$, then

$$\frac{1}{\sqrt{n}} M_n \xrightarrow{d} W,$$

where W is Brownian motion in time A.

True, that not knowing y we do not know A, but we will address this problem in appropriate places below.

The model in which the compensator of the point process of interest has a derivative, which is a product of a random process, that is Y_n here, and a deterministic function, which is μ here, is called a multiplicative model or Aalen's model (see e.g. Andersen et al. [1993]).

11.1 Testing a simple hypothesis.

Let us consider how can we use the limit theorem (10.4) in testing a simple hypothesis that the unknown distribution function F is actually equal to a given, or prescribed, distribution function F_0, $F = F_0$. Basically, we do it in the same way as we used the limit theorem for empirical processes in Lecture 5. Now, however, we need to clarify an additional question, which is this: the limit behavior of $\frac{1}{\sqrt{n}}M_n$, and, hence, the statistics based on it, depends on the distribution function G of the censoring variables, which we do not know.

Several ways out of this difficulty can be suggested.

One is that we can use (10.6) as an estimator of the unknown function A (cf. with (4.3.6) and (4.3.18) in Andersen et al. [1993]). This works well, for example, when one uses the chi-square goodness of fit test. If $0 = t_0 < t_1 < \ldots < t_k$ are selected boundary points for age intervals, then the following statistic, cf. with (6.5),

$$\sum_{j=0}^{k} \frac{\left[\Delta N_n(t_j) - \int_{t_j}^{t_{j+1}} Y_n(s)\mu_0(s)\,ds \right]^2}{\int_{t_j}^{t_{j+1}} Y_n(s)\mu_0(s)\,ds}, \tag{11.3}$$

where, as usual, $\Delta N_n(t_j) = N_n(t_{j+1}) - N_n(t_j)$, and $\mu_0(s) = f_0(s)/[1 - F_0(s)]$ is the hypothetical force of mortality, will have a chi-square limit distribution with $k+1$ degrees of freedom if the hypothesis $F = F_0$ is true.

◇ **Exercise.** Verify this statement using (10.4). And why, indeed, do we have $k+1$ degrees of freedom and not k, as in Lectures 5 and 6? △

In the case of the Kolmogorov–Smirnov goodness of fit statistic it is easier to apply the following consideration: it is clear that

$$\sup_{0 \leq t < \infty} |W(t)| = \sup_{0 \leq t < \infty} |w(A(t))| = \sup_{0 \leq x < A(\infty)} |w(x)|$$

so that the distribution on the right-hand side depends not on the whole function $A(t)$, $t > 0$, but only on $A(\infty) = \int_0^\infty [1 - G(s)]\,dF(s) = P\{Y_i >$

T_i}, and it is easy to estimate this probability by the following frequency

$$\widehat{A}_n = \frac{1}{n}\sum_{i=1}^{n}\delta_i = \frac{1}{n}\sum_{i=1}^{n}I_{\{Y_i>T_i\}}.$$

It is obvious that $\widehat{A}_n \to A(\infty)$ with probability 1 when $n \to \infty$, and therefore,

$$P\left\{\sup_{0\leq x<\widehat{A}_n}|w(x)|>\lambda\right\} \longrightarrow P\left\{\sup_{0\leq x<A(\infty)}|w(x)|>\lambda\right\}, \quad n\to\infty.$$
$$(11.4)$$

Then we can use, for sufficiently large n, the known probability on the left as an approximation for the probability on the right. In more detail, for each finite A, the processes $w(x)$ and $\sqrt{A}\,w(x/A)$ have the same distribution. Therefore statistics

$$\sup_{0<x<A}|w(x)| \quad \text{and} \quad \sqrt{A}\sup_{0<y<1}|w(y)|$$

also have the same distribution, and as an approximation of the distribution of the Kolmogorov–Smirnov statistic we obtain

$$P\left\{\sup_{0<t<\infty}\left|N_n(t) - \int_0^t Y_n(s)\mu_0(s)\,ds\right| > \lambda\sqrt{n}\right\}$$
$$-P\left\{\sup_{0<y<1}|w(y)| > \lambda/\sqrt{\widehat{A}_n}\right\} \longrightarrow 0, \quad n\to\infty. \quad (11.5)$$

◇ **Exercise.** Consider a rather elegant third possibility, which is to use instead of M_n the weighted process

$$\eta_n(t) = \int_0^t \frac{1}{Y_n^{1/2}(s)}M_n(ds)$$
$$= \int_0^t \left[\frac{dN_n(s)}{Y_n^{1/2}(s)} - Y_n^{1/2}(s)\mu_0(s)\,ds\right],$$

where we assume $Y_n^{-1/2}(s)M_n(ds) = 0$ if $Y_n^{1/2}(s) = 0$, that is, if

$s > \tilde{T}_{(n)}$. Verify that this process is again a martingale, and that its quadratic variation does not depend on Y_n as such but only on $\tilde{T}_{(n)}$:

$$\langle \eta_n \rangle (t) = \int_0^t \mu_0(s)\, ds = -\ln(1 - F_0(t)), \qquad t \leq \tilde{T}_{(n)}.$$

Try to show that for any deterministic moment τ' such that $[1 - F_0(\tau')][1 - G(\tau')] > 0$, the process $\eta_n(t), 0 < t \leq \tau'$, converges in distribution to the process W,

$$W(t) = w(-\ln(1 - F_0(t))).$$

As we can see, goodness of fit statistics based on the process $\eta_n(t)$, $0 < t \leq \tau'$, are free from the unknown distribution function G. $\triangle$

◇ **Exercise.** Using computer simulation experiments investigate how good the convergence

$$\sup_{0 < t < \tau'} |\eta_n(t)| \overset{d}{\to} \sup_{0 < x < -\ln[1 - F_0(\tau')]} |w(x)|$$

is for various F_0 (see Lecture 2). Is this approximation satisfactory already for $n \approx 100$? $\triangle$

11.2   Testing a parametric hypothesis

In Lecture 7 we had a chance to see that testing of a simple hypothesis $F = F_0$, that is, the hypothesis where F is completely fixed, does not occur in practice too often. Much more frequently one has to consider parametric hypotheses, where the hypothetical distribution function and, therefore, the hypothetical force of mortality depend on a parameter, or parameters; and their values are unknown and have to be estimated from the observations at hand.

Namely, they choose a family $\{F_\theta\}$ of distribution functions, depending on a parameter θ, and the corresponding family $\{\mu_\theta\}$, $\mu_\theta = f_\theta/(1 - F_\theta)$, and the hypothesis to be tested is

H: There exists a value of θ, such that $\mu(x) = \mu(x, \theta)$,   (11.6)

but what this value of θ is the hypothesis does not prescribe – it remains unknown and we replace it in our testing procedures by its estimator $\widehat{\theta}$. As a consequence, the process

$$M_n(t,\widehat{\theta}) = N_n(t) - \int_0^t Y_n(y)\mu(y,\widehat{\theta})\,dy,$$

unlike the process (9.6), is not a martingale any more and the limit theorem (10.4) for $M_n(\cdot,\widehat{\theta})$ is not true.

The possibilities to resolve this problem were investigated in Maglapheridze et al. [1989] in a relatively general framework. Namely, using the method suggested in Khmaladze [1981], the authors proposed a transformation of $M_n(\cdot,\widehat{\theta})$, similar to the one which we used in Lecture 7, and proved that the limit theorem (10.4) for this transformation is again true. The transformation is one-to-one, which, informally, means that "all statistical information", contained in the process $M_n(\cdot,\widehat{\theta})$, is also contained in the transformed process $W_n(\cdot,\widehat{\theta})$, while the asymptotic behavior of $W_n(\cdot,\widehat{\theta})$ is much simpler.

Let us derive the form of $W_n(\cdot,\widehat{\theta})$, somewhat simplifying the approach of Maglapheridze et al. [1989] on the way. In its key features $W_n(\cdot,\widehat{\theta})$ is essentially no different from the construction of the process W_n of Lecture 7, and is somewhat simpler. For clarity, assume that θ is one-dimensional and denote

$$h(t,\theta) = \frac{\partial}{\partial\theta}\ln\mu(t,\theta) \quad \text{and} \quad A_n(t,\theta) = \int_0^t \frac{Y_n(s)}{n}\mu(s,\theta)\,ds. \quad (11.7)$$

It will be useful to note that, assuming a simple differentiability property of $\mu(t,\theta)$ in θ, we can ascertain that in a Taylor expansion in θ

$$\frac{1}{\sqrt{n}}\left(M_n(t,\widehat{\theta}) - M_n(t,\theta)\right) = \int_0^t h(t,\theta)dA_n(s)\sqrt{n}(\widehat{\theta} - \theta) + r_n(t,\widehat{\theta},\theta)$$

$$(11.8a)$$

the remainder term is asymptotically small uniformly in t:

$$\sup_t |r_n(t,\widehat{\theta},\theta)| \to 0, \quad n \to \infty. \quad (11.8b)$$

As in Lecture 7, we again need a linear regression of $dN_n(t)$, this time – only on $\int_t^\tau h(s,\theta)dN_n(s)$. The easiest way to derive the coefficient of this linear regression is to use the general formula: if $g'(s)$ and $g(s)$ are two deterministic functions, then the covariance between the two integrals $\int_0^\tau g'(s)dM_n(s,\theta)$ and $\int_0^\tau g(s)dM_n(s,\theta)$ is given by

$$\mathsf{E}\left(\int_0^\tau g'(s)dM_n(s,\theta)\int_0^\tau g(s)dM_n(s,\theta)\right)=n\mathsf{E}\int_0^\tau g'(s)g(s)dA_n(s,\theta).$$

(11.9)

We discuss this formula at the end of this lecture. Right now let us apply it, with $g'(s)=I_{\{t\leq s<t+dt\}}$ and $g(s)=h(s,\theta)I_{\{t\leq s<\infty\}}$ with a fixed t, to obtain

$$\mathsf{E}\left(dM_n(t,\theta)\int_t^\tau h(s,\theta)dM_n(s,\theta)\right)=n\mathsf{E}h(t,\theta)dA_n(t,\theta)$$

and

$$\mathsf{E}\left(\int_t^\tau h(s,\theta)dM_n(s,\theta)\right)^2=n\mathsf{E}\int_s^\infty h^2(s,\theta)dA_n(s,\theta).$$

From this we find that the coefficient of the linear regression is

$$a=\frac{\mathsf{E}h(t,\theta)dA_n(t,\theta)}{\mathsf{E}\int_t^\infty h^2(s,\theta)dA_n(s,\theta)},$$

and thus the difference

$$dM_n(t,\theta)-\frac{\mathsf{E}h(t,\theta)dA_n(t,\theta)}{\mathsf{E}\int_t^\infty h^2(s,\theta)dA_n(s,\theta)}\int_t^\tau h(s,\theta)dM_n(s,\theta)$$

is uncorrelated with all $M_n(s,\theta), s\leq t$, and $\int_t^\tau h(s,\theta)dM_n(s,\theta)$. In other words, with respect to $\{\mathscr{F}_t, t\geq 0\}$, where each σ-algebra is defined as

$$\mathscr{F}_t=\sigma\{M_n(s,\theta), s\leq t, \int_0^\tau h(s,\theta)dM_n(s,\theta)\},$$

the process

$$\frac{1}{\sqrt{n}}\left[M_n(t,\theta)-\int_0^t\frac{\mathsf{E}h(s,\theta)dA_n(s,\theta)}{\mathsf{E}\int_s^\infty h^2(y,\theta)dA_n(y,\theta)}\int_s^\tau h(y,\theta)dM_n(y,\theta)\right]$$

is the process with uncorrelated increments.

If $Y_n(t)/n$ converges to a deterministic function, the approximation

$$a \approx \frac{h(t,\theta)dA_n(t,\theta)}{\int_t^\infty h^2(s,\theta)dA_n(s,\theta)}$$

can be justified. Then the resulting difference

$$M_n(t,\theta) - \int_0^t \frac{h(s,\theta)dA_n(s,\theta)}{\int_s^\infty h^2(s,\theta)dA_n(s,\theta)} \int_s^\tau h(y,\theta)dM_n(y,\theta)$$

will be asymptotically a process with uncorrelated increments.

So far we have used $M_n(t,\theta)$. Now we need to go back and use $M_n(t,\widehat{\theta})$ in its place. The key property is, however, that the change in the resulting process, the difference above, will be asymptotically negligible.

Indeed, if we replace $M_n(s,\theta)$ in the last display by the integral in the leading term in (11.8), the result will be zero:

$$h(t,\theta)dA_n(t) - \frac{h(t,\theta)dA_n(t,\theta)}{\int_t^\infty h^2(s,\theta)dA_n(s,\theta)} \int_t^\tau h^2(s,\theta)dA_n(s) = 0. \quad (11.10)$$

Assume in addition to (11.8) that

$$\sup_t \left| \int_t^\tau h(s,\theta)dr_n(s,\widehat{\theta},\theta) \right| \to 0.$$

Then eventually we obtain

$$W_n(t,\widehat{\theta}) = \frac{1}{\sqrt{n}} \left[M_n(t,\widehat{\theta}) \right.$$
$$\left. - \int_0^t \frac{h(s,\widehat{\theta})}{\int_s^\infty h^2(y,\widehat{\theta})dA_n(y,\theta)} \int_s^\tau h(y,\widehat{\theta})M_n(dy,\widehat{\theta})dA_n(s,\widehat{\theta}) \right]$$

and the trajectories of this process can be calculated.

For the process $W_n(\cdot,\widehat{\theta})$ the following limit theorem (see Maglapheridze et al. [1989]), is true:

if the hypothesis (11.6) is true and if $W = w \circ A$ is a Brownian motion defined in (10.4) and (10.5), then

$$W_n(\cdot,\widehat{\theta}) \xrightarrow{d} W. \quad (11.11)$$

◇ **Exercise.** If the parameter θ is multi-dimensional, then the function $h(t, \theta)$ will be a vector function and $\int_s^\infty h^2(y, \widehat{\theta}) dA_n(y, \theta)$ will be a matrix. Re-write the expression for $W_n(t, \widehat{\theta})$ in this case. Verify that the analogue of (11.10) is again true.   △

As a corollary to this theorem, for the Kolmogorov–Smirnov statistic

$$\sup_t |W_n(t, \widehat{\theta})|$$

as well as for the chi-square statistic

$$\sum_{j=0}^k \frac{[\Delta W_n(t_j, \widehat{\theta})]^2}{\Delta A_n(t_j)}$$

everything that we said in the previous part of this lecture about these statistics from M_n, see (11.3) and (11.5), is fully applicable.

As an example, let us apply our formulae to the practically interesting case of the Gompertz distribution, introduced in (2.5) and (2.6) with

$$\mu(t, \theta_1, \theta_2) = \theta_1 \theta_2 e^{\theta_2 t}.$$

If the unknown parameter is the "shape" parameter θ_2, then

$$h(s, \theta) = \left(\frac{1}{\theta_2} + s\right) \quad \text{and} \quad dA_n(s) = \frac{Y_n(s)}{n} \theta_1 \theta_2 e^{\theta_2 s} ds.$$

Substitution of these expressions in the formula for $W_n(t, \widehat{\theta})$ gives the desired process.

It is interesting to see what happens in the case when the unknown parameter is the θ_1. In this case $h(s, \theta) = 1/\theta_1$, and it cancels out in the integral term in $W_n(t, \widehat{\theta})$. So $W_n(t, \widehat{\theta})$ becomes particularly simple:

$$W_n(t, \widehat{\theta}) = \frac{1}{\sqrt{n}} \left[M_n(t, \widehat{\theta}) - \int_0^t \frac{M_n(\tau, \widehat{\theta}) - M_n(s, \widehat{\theta})}{A_n(\tau, \widehat{\theta}) - A_n(s, \widehat{\theta})} dA_n(s, \widehat{\theta}) \right].$$

The reader probably noticed that in the limit theorem (11.11) we did not need to specify an asymptotic form of the estimator $\widehat{\theta}$. However, if the estimator of θ_1, $\widehat{\theta}_1$ was obtained from the equation

$$M_n(\tau, \widehat{\theta}_1) = 0,$$

then the process

$$\frac{1}{\sqrt{n}} M_n(t, \widehat{\theta}_1)$$

becomes asymptotically very convenient: if V is a Brownian bridge in time $F(t) = A(t)/A(\tau)$ then

$$\frac{1}{\sqrt{n}} M_n(\cdot, \widehat{\theta}_1) \xrightarrow{d} \sqrt{A(\tau)}\, V, \tag{11.12}$$

and, therefore, we know the limit distribution of wide class of test statistics based on $\frac{1}{\sqrt{n}} M_n(\cdot, \widehat{\theta}_1)$. The proof of this fact we formulate as an exercise.

◇ **Exercise.** Verify the following steps:
  a) the estimator $\widehat{\theta}_1$ has asymptotic representation

$$\sqrt{n}(\widehat{\theta}_1 - \theta_1) = \frac{1}{\sqrt{n}} \frac{\theta_1 M_n(\tau, \widehat{\theta}_1)}{A_n(\tau)} + o_P(1);$$

  b) therefore

$$\frac{1}{\sqrt{n}} M_n(\cdot, \widehat{\theta}_1) = \frac{1}{\sqrt{n}} \left[M_n(\cdot, \theta_1) - \frac{A_n(t)}{A_n(\tau)} M_n(\tau, \theta_1) \right];$$

  c) if W is a Brownian bridge in time $F(t) = A(t)/A(\tau)$, then the process on the interval $[0, \tau)$

$$V(t) = W(t) - F(t)W(\tau)$$

is a Brownian bridge on the same interval.
Use now the limit theorem for $M_n(\cdot, \theta)$ and conclude that the limit theorem (11.12) is correct.   △

The fact uncovered by (11.12) can be generalized: if the parameter

θ is one-dimensional and its estimator $\widehat{\theta}$ is chosen as the root of the equation

$$\int_0^{\widetilde{T}_{(n)}} h(s,\theta)dM_n(s,\theta) = 0,$$

which is the maximum likelihood equation, then for the process

$$\xi_n(t) = \frac{1}{(\int_0^\tau h^2(s,\widehat{\theta})A_n(ds,\widehat{\theta}))^{1/2}} \int_0^t h(s,\widehat{\theta})M_n(ds,\widehat{\theta})$$

the limit statement

$$\xi_n \xrightarrow{d} V = v \circ F,$$

with

$$F(t) = \int_0^t h^2(s,\theta)A(ds,\theta) \Big/ \int_0^\infty h^2(s,\theta)A(ds,\theta),$$

is true.

Therefore, one can derive statistical inference about F in exactly the same way as we have seen it in Lecture 5.

What happens if both parameters are unknown and have to be estimated? As in the general case, we start with linear regression of $dM_n(t,\widehat{\theta})$ on

$$\int_t^\tau h(s,\theta)dM_n(s,\theta) = \begin{pmatrix} \frac{1}{\theta_1}[M_n(\tau,\theta) - M_n(t,\theta)] \\ \int_t^\tau (\frac{1}{\theta_2} + s)dM_n(s,\theta) \end{pmatrix}.$$

We should recall, however, that "regression on a vector" is a short phrase for, or is the same as, regression on the linear subspace generated by this vector. Therefore our regression on $\int_t^\tau h(s,\theta)dM_n(s,\theta)$ is the same as regression on

$$\begin{pmatrix} M_n(\tau,\theta) - M_n(t,\theta) \\ \int_t^\tau sdM_n(s,\theta) \end{pmatrix}.$$

Moreover, the second coordinate we can "orthogonalize" with respect to the first and consider

$$\int_t^\tau (s - \frac{\int_t^\tau sdA_n(s)}{A_n(\tau) - A_n(t)})dM_n(s,\theta)$$

instead. As it follows from (11.9), this integral and $M_n(\tau, \theta) - M_n(t, \theta)$ are uncorrelated. Therefore, coefficients of the linear regression

$$a_t[M_n(\tau, \theta) - M_n(t, \theta)] + b_t \int_t^\tau \left(s - \frac{\int_t^\tau s \, dA_n(s)}{A_n(\tau) - A_n(t)}\right) dM_n(s, \theta)$$

can be found independently from each other. In particular,

$$a_t = \frac{E \, dM_n(t, \theta)[M_n(\tau, \theta) - M_n(t, \theta)]}{\text{Var}[M_n(\tau, \theta) - M_n(t, \theta)]}$$

$$= \frac{E \, dA_n(t)}{E[A_n(\tau) - A_n(t)]}.$$

Again, if Y_n/n converges to a deterministic function we can obtain that

$$a_t \approx \frac{dA_n(t)}{A_n(\tau) - A_n(t)} = \frac{Y_n(t)e^{\theta_2 t}}{\int_t^\tau Y_n(s)e^{\theta_2 s} ds} dt.$$

Similarly, with

$$c(t) = \frac{\int_t^\tau s \, dA_n(s)}{A_n(\tau) - A_n(s)} = \frac{\int_t^\tau s \, Y_n(s)e^{\theta_2 s} ds}{\int_t^\tau Y_n(s)e^{\theta_2 s} ds},$$

we obtain

$$b_t = \frac{E \, dM_n(t, \theta) \int_t^\tau (s - c(t)) dM_n(s, \theta))}{\text{Var} \int_t^\tau (s - c(t)) dM_n(s, \theta)}$$

$$= \frac{E(t - c(t)) dA_n(t)}{E \int_t^\tau [s - c(t)]^2 dA_n(t)}$$

and this, again, can be simplified to

$$b_t \approx \frac{[t - c(t)] Y_n(t)e^{\theta_2 t}}{\int_t^\tau [s - c(t)]^2 Y_n(s)e^{\theta_2 s} ds} dt.$$

◇ **Exercise.** Find simplifications that will facilitate numerical calculations. For example, verify, using integration by parts, that all integrals in the eventual expression for a_t and b_t are actually sums, and note that in $c(t)$

$$\int_t^\tau s \, Y_n(s)e^{\theta_2 s} ds = \frac{\partial}{\partial \theta_2} \int_t^\tau Y_n(s)e^{\theta_2 s} ds.$$

△

11.3 Two-sample problem for censored observations

In this form of the two-sample problem they consider two indepen-
dents samples, and observations in each of them are subject to possible
random censoring. In other words, there are two sequences of pairs
of random variables $\{(T_{1j}, Y_{1j})\}_{j=1}^{n}$ and $\{(T_{2j}, Y_{2j})\}_{j=1}^{m}$, of, generally,
different lengths n and m, where T_{1j}, $j = 1, \ldots, n$, are lifetimes of individuals from the first cohort and Y_{1j}, $j = 1, \ldots, n$, are the censoring
variables, while T_{2j}, $j = 1, \ldots, m$, are lifetimes of the individuals from
the second cohort and $Y_{2j}, j = 1, \ldots, m$, are the respective censoring
variables. Due to censorship, we can only observe $\widetilde{T}_{ij} = \min(T_{ij}, Y_{ij})$
and $\delta_{ij} = I_{\{T_{ij} = \widetilde{T}_{ij}\}}$. In other words, we observe two samples

$$(\widetilde{T}_{11}, \delta_{11}), (\widetilde{T}_{12}, \delta_{12}), \ldots, (\widetilde{T}_{1n}, \delta_{1n})$$

and

$$(\widetilde{T}_{21}, \delta_{21}), (\widetilde{T}_{22}, \delta_{22}), \ldots, (\widetilde{T}_{2m}, \delta_{2m}).$$

Denote by F_1 and G_1 distribution functions of T_{1j} and Y_{1j}, while F_2 and
G_2 denote distribution functions of T_{2j} and Y_{2j}, respectively. We wish
to test that

$$F_1 = F_2, \tag{11.13}$$

stating nothing about distributions of the censoring variables. In other
words, we wish to test that partially observable samples $\{T_{1j}\}_{j=1}^{n}$ and
$\{T_{2j}\}_{j=1}^{n}$ have been homogeneous.

Let us stress that testing homogeneity is a very important statistical
problem both in demography and in insurance business.

For example, will three randomly selected individuals of the same
gender, the same age, similar in health parameters, non-smoking and
living in the same town, but such that one bought an annuity, another
bought life insurance and the third did neither, have the same distribution function of lifetime? A lay person, who has never considered
such questions, may well say that yes, they probably do. But an actuary may say that no, most likely, the distribution functions will be
different. It is not important for us here why (or if) this is happening
(see the remark on mixtures in Lecture 12 for more comments). Instead
we rather point out another aspect of the two-sample problem. It often

looks useful to amalgamate several portfolios into one, but in doing this one would need to understand whether the amalgamated portfolios were homogeneous. Conversely, it may also be useful to subdivide one large inhomogeneous portfolio into several ones and require different premiums in each. But for this again, one needs to understand how prudently the hypothesis of homogeneity is rejected.

Let $N_{in}(t)$, $Y_{in}(t)$, $M_{in}(t)$, $t \geq 0$, $i = 1, 2$, denote the processes considered in Lecture 8 and based on our first and second samples, respectively. It is not difficult to see that

$$\widehat{\Lambda}_{1n}(t) = \int_0^t \frac{dN_{1n}(s)}{Y_{1n}(s)} \qquad \widehat{\Lambda}_{2n}(t) = \int_0^t \frac{dN_{2m}(s)}{Y_{2m}(s)},$$

are unbiased estimations of the risk functions, see (1.15), and therefore, if the hypothesis (11.13) is true, the expected value of the difference

$$\int_0^t \left[\frac{dN_{1n}(s)}{Y_{1n}(s)} - \frac{dN_{2m}(s)}{Y_{2m}(s)} \right]$$

is 0. Moreover, this difference can be re-written as

$$\int_0^t \left[\frac{dM_{1n}(s)}{Y_{1n}(s)} - \frac{dM_{2m}(s)}{Y_{2m}(s)} \right] + \Lambda_1(t) - \Lambda_2(t),$$

and this shows that if (11.13) is true, our difference is a martingale with respect to the "pooled" filtration $\{\mathscr{F}_t^{m,n}, t \geq 0\}$, where

$$\mathscr{F}_t^{m,n} = \mathscr{F}_{1t}^n \vee \mathscr{F}_{2t}^m = \sigma\{N_{1n}(s), Y_{1n}(s), N_{2m}(s), Y_{2m}(s), s \leq t\}.$$

For reasons that will soon become clear, it is convenient to use (random) weight function $K(s)$, which we will choose as a combination of $Y_{1n}(s)$ and $Y_{2m}(s)$, see (11.15) and (11.16), and consider a process

$$\int_0^t K(s) \left[\frac{dN_{1n}(s)}{Y_{1n}(s)} - \frac{dN_{2m}(s)}{Y_{2m}(s)} \right]$$

$$= M_{n,m}(t) + \int_0^t K(s) [d\Lambda_1(s) - d\Lambda_2(s)], \qquad (11.14)$$

where

$$M_{n,m}(t) = \int_0^t K(s) \left[\frac{dM_{1n}(s)}{Y_{1n}(s)} - \frac{dM_{2m}(s)}{Y_{2m}(s)} \right]$$

is a martingale with respect to $\{\mathscr{F}_t^{m,n}, t \geq 0\}$. The quadratic variation of this martingale is equal to

$$\langle M_{m,n} \rangle (t) = \int_0^t K^2(s) \left[\frac{1}{Y_{1n}^2(s)} \langle M_{1n} \rangle (ds) + \frac{1}{Y_{2m}^2(s)} \langle M_{2m} \rangle (ds) \right].$$

Using the equality (10.3) one can re-write the last expression as

$$\langle M_{m,n} \rangle (t) = \int_0^t K^2(s) \frac{Y_{1n}(s) + Y_{2m}(s)}{Y_{1n}(s) Y_{2m}(s)} \mu(s) \, ds.$$

This expression involves, as we can see, the common, but unknown to us, rate of mortality μ. However, it is natural to replace

$$[Y_{1n}(s) + Y_{2m}(s)] \mu(s) \, ds$$

by

$$dN(s) = dN_{1n}(s) + dN_{2m}(s)$$

and thus use the computationally even simpler expression

$$A^{m,n}(t) = \int_0^t K^2(s) \frac{1}{Y_{1n}(s) Y_{2m}(s)} \, dN(s).$$

If now, for this or that choice of K, we can establish convergence of $\langle M_{m,n} \rangle$ to a limit and verify Lindeberg's condition, then under hypothesis (11.13) the left-hand side of (11.14) will converge in distribution to an appropriate Brownian motion W, and statistics from it – to statistics from this W. If the hypothesis is not true, then the resulting shift will change the limit distribution of our statistics, sometimes substantially, and this is what we will notice.

As was done in Gill [1980], with references to the papers Gehan [1965] and Cox [1972], consider the following choices of the weight function K:

$$K_G(s) = \frac{1}{\sqrt{mn(m+n)}} Y_{1n}(s) Y_{2m}(s),$$

$$K_C(s) = \sqrt{\frac{m+n}{mn}} \frac{Y_{1n}(s) Y_{2m}(s)}{Y_{1n}(s) + Y_{2m}(s)} \tag{11.15}$$

to which we add here one more choice of

$$K(s) = \sqrt{\frac{Y_{1n}(s) Y_{2m}(s)}{m+n}} \quad \text{and} \quad K(s) = \sqrt{\frac{Y_{1n}(s) Y_{2m}(s)}{Y_{1n}(s) + Y_{2m}(s)}}. \tag{11.16}$$

Under the first choice, the martingale $M^{m,n}$ becomes

$$M_G^{m,n}(t) = \frac{1}{\sqrt{mn(m+n)}} \int_0^t [Y_{2m}(s)\,dN_{1n}(s) - Y_{1n}(s)\,dN_{2m}(s)],$$

and its quadratic variation becomes

$$\langle M_G^{m,n} \rangle(t) = \int_0^t \frac{Y_{1n}(s)}{n}\,\frac{Y_{2m}(s)}{m}\,\frac{Y_{1n}(s)+Y_{2m}(s)}{m+n}\,\mu(s)\,ds.$$

We approximate this quadratic variation by

$$A_G^{m,n}(t) = \int_0^t \frac{Y_{1n}(s)}{n}\,\frac{Y_{2m}(s)}{m}\,dN(s),$$

which is also computationally simpler – cf. Breslow and Crowley [1970], where $A_G^{m,n}(\infty)$ was considered.

To verify the conditions of the central limit theorem for $M_G^{m,n}$ is easy. The resulting statement is

   if the hypothesis (11.13) is true and $m,n \to \infty$, then

$$M_G^{m,n} \xrightarrow{d} W = w \circ A_G,$$

where

$$A_G(t) = \int_0^t [1 - G_1(s)][1 - G_2(s)][2 - G_1(s) - G_2(s)][1 - F(s)]^2\,dF(s),$$

and where F denotes the common distribution function of T_{1j} and T_{2j}.

◇ **Exercise.** a) Verify the conditions of Rebolledo martingale central limit theorem (Lecture 8) for $M_G^{m,n}$ and obtain an expression for A_G.

b) If the hypothesis (11.13) is not true, what will change in the limit behavior of $M_G^{m,n}$?

c) Is it really possible to replace $[Y_{1n}(s)+Y_{2m}(s)]\mu(s)\,ds$ by $dN(s)$? In other words, is the difference

$$\frac{1}{m+n}\left[N(t) - \int_0^t [Y_{1n}(s)+Y_{2m}(s)]\mu(s)\,ds\right]$$

asymptotically negligible?   △

From the previous considerations it follows that if the hypothesis (11.13) is true, then for large m and n for the Kolmgorov–Smirnov statistic we obtain

$$P\left\{\sup_{0\le t<\infty} |M_G^{m,n}(t)| > \lambda\right\} \approx P\left\{\sup_{0<y<1} |w(y)| > \lambda \Big/ \sqrt{A_G^{m,n}}\right\},$$

where $A_G^{m,n}$ denotes the value of $A_G^{m,n}(t)$ in the last point:

$$A_G^{m,n} = A_G^{m,n}\left(\min(\widetilde{T}_{1(n)}, \widetilde{T}_{2(m)})\right).$$

For the two-sample chi-square statistic

$$X_{k+1}^2(m,n) = \sum_{j=0}^{k} \frac{\left(\int_{t_j}^{t_{j+1}} [Y_{2m}(s)\,dN_{1n}(s) - Y_{1n}(s)\,dN_{2m}(s)]\right)^2}{\Delta A_G^{m,n}(t_j)},$$

where again $0 = t_0 < t_1 < \cdots < t_k$ are the boundaries of the class-intervals, we have that if $m,n \to \infty$ then the limit distribution of this statistics is a chi-square distribution with $k+1$ degrees of freedom.

With the second choice of K our process becomes

$$M_C^{m,n}(t) =$$
$$\sqrt{\frac{m+n}{mn}} \int_0^t \frac{1}{Y_{1n}(s)+Y_{2m}(s)} \left[Y_{2m}(s)dN_{1n}(s) - Y_{1n}(s)dN_{2m}(s)\right],$$

and its quadratic variation becomes

$$\langle M_C^{m,n}\rangle(t) = \frac{m+n}{mn} \int_0^t \frac{Y_{1n}(s)\,Y_{2m}(s)}{Y_{1n}(s) + Y_{2m}(s)}\,\mu(s)\,ds.$$

Again, in view of the unknown μ, it can be approximated by

$$A_C^{m,n}(t) = \frac{m+n}{mn} \int_0^t \frac{Y_{1n}(s)\,Y_{2m}(s)}{[Y_{1n}(s) + Y_{2m}(s)]^2}\,N(ds).$$

All the rest, concerning the Kolmgorov–Smirnov statistic and two-sample chi-square statistic, based on the process $M_C^{m,n}$, stays without change.

The choice of K suggested in (11.16) offers a clear illustration of how one can vary the limiting function A, the limiting "time", and how simple it can be made by varying the weight function. Indeed, with the first choice of K from (11.16) we obtain

$$M^{mn}(t) = \frac{1}{\sqrt{m+n}} \int_0^t \left[\sqrt{\frac{Y_{2m}(s)}{Y_{1n}(s)}} dN_{1n}(s) - \sqrt{\frac{Y_{1n}(s)}{Y_{2m}(s)}} dN_{2m}(s) \right]$$

with

$$\langle M^{m,n} \rangle(t) = \frac{1}{m+n} \int_0^t \left[Y_{1n}(s) + Y_{2m}(s) \right] \mu(s) \, ds,$$

which can be directly approximated simply by

$$\frac{1}{m+n} N(t).$$

With the second choice of K from (11.16) we have

$$M^{mn}(t) = \int_0^t \frac{1}{\sqrt{Y_{1n}(s) + Y_{2m}(s)}}$$

$$\times \left[\sqrt{\frac{Y_{2m}(s)}{Y_{1n}(s)}} dN_{1n}(s) - \sqrt{\frac{Y_{1n}(s)}{Y_{2m}(s)}} dN_{2m}(s) \right]$$

with even simpler quadratic variation

$$\langle M^{m,n} \rangle(t) = \int_0^t \mu(s) \, ds, \qquad t \leq \min\left(\tilde{T}_{1(n)}, \tilde{T}_{2(m)} \right).$$

It is worth noting that in deriving the expressions and approximations for quadratic variations of our processes we avoided any need for knowing distribution functions G_1 and G_2 of the censoring random variables. Therefore, although we devoted some space to this question, the choice of weight function is, as a matter of fact, not that important, especially since we did not discuss any specific class of alternative distributions. In practice, users can choose the weight functions, which in their opinion seems simplest.

11.4 Why is (11.9) **correct? Ito integral**

The equality (11.9) visually is very similar to (10.13), but there is one essential difference in the construction when g and g' are random. In particular, we need to understand what sort of randomness g and g' can have to allow for fruitful theory. The situation will be particularly visible when functions g and g' are step-wise constant.

Let us skip θ as inessential in this section. The calculation of the integral

$$\int_0^\tau g(s)dM_n(s) = \int_0^\tau g(s)dN_n(s) - \int_0^\tau g(s)Y_n(s)\mu(s)ds$$

when $g(s)$ is random does not require much of a structure from g. For example, if trajectories of g are continuous, the integral can simply be understood as a Stieltjes integral, for each trajectory of g, N_n and Y_n. However, to be able to have a systematic calculus of such integrals we need a specific property from $g(t)$. Namely, we assume that $g(t)$ is continuous from the left, and, as random variable, depends on the values of $N_n(s), Y_n(s)$ for $s < t$ only. That is, if the trajectories of $N_n(s)$ and $Y_n(s)$ for $s < t$ are fixed – recall that these trajectories form a "point" in σ-algebra $\mathscr{F}_t$ – then $g(t)$ is also fixed. For instance, $g(t) = (N_n(\tau)/\tau)t$ or $g(t) = N_n(\tau)t$ does not have this property, because at each t it incorporates some amount of the "future" in the form of $N_n(\tau) - N_n(t-)$, while $g(t) = Y_n(t)\mu(t)$ does have it. The function g with the above property is predictable: its value at each t can be exactly predicted by the history up to, but not including, this t. Therefore, obviously,

$$E[g(t)|\mathscr{F}_t] = g(t).$$

Given a partition $0 = t_0 < t_1 < \cdots < t_{n+1} = \tau$, suppose g has the property above, and, in addition, is piece-wise constant. Then define the integral

$$\int_0^\tau g(s)dM_n(s) = \sum_{i=0}^n g(t_i)\left[M_n(t_{i+1}) - M_n(t_i)\right]$$

. Note that we did not take the values of g anywhere inside $[t_i, t_{i+1}]$ but exactly at its left end-point. This is essential, and the properties of the integral will be altered if we use $g(\tilde{t}_i), t_i < \tilde{t}_i \le t_{i+1}$.

Now suppose g' has the same properties as g and is also step-wise constant on the same grid. Consider the expected value

$$E \int_0^\tau g(s)dM_n(s) \int_0^\tau g'(s)dM_n(s)$$

$$= E \sum_{i=0}^n \sum_{j=0}^n g(t_i)g'(t_j) \left[M_n(t_{i+1}) - M_n(t_i)\right] \left[M_n(t_{j+1}) - M_n(t_j)\right].$$

In the double sum consider first the summands with $i \neq j$. Assume, say, that $i < j$. Then $t_{i+1} \leq t_j$. Take conditional expectation inside the unconditional expectation:

$$Eg(t_i)g'(t_j) \left[M_n(t_{i+1}) - M_n(t_i)\right] \left[M_n(t_{j+1}) - M_n(t_j)\right]$$
$$= EE[g(t_i)g'(t_j) \left[M_n(t_{i+1}) - M_n(t_i)\right] \left[M_n(t_{j+1}) - M_n(t_j)\right] | \mathscr{F}_{t_j}].$$

However, given $\mathscr{F}_{t_j}$ all three of $g(t_i), g'(t_j)$ and $\left[M_n(t_{i+1}) - M_n(t_i)\right]$ are constants. Therefore they can be taken outside the conditional expectation. But then

$$Eg(t_i)g'(t_j) \left[M_n(t_{i+1}) - M_n(t_i)\right] E\left[\left[M_n(t_{j+1}) - M_n(t_j)\right] | \mathscr{F}_{t_j}\right] = 0$$

because

$$E\left[\left[M_n(t_{j+1}) - M_n(t_j)\right] | \mathscr{F}_{t_j}\right] = 0.$$

Hence, only the terms with $i = j$ remain, and

$$EE[g(t_i)g'(t_i) \left[M_n(t_{i+1}) - M_n(t_i)\right]^2 | \mathscr{F}_{t_i}]$$
$$= Eg(t_i)g'(t_i) \int_{t_i}^{t_{i+1}} Y_n(s)\mu(s)ds.$$

Summation over i leads to

$$E \int_0^\tau g(s)dM_n(s) \int_0^\tau g'(s)dM_n(s) = E \sum_{i=0}^n g(t_i)g'(t_i) \int_{t_i}^{t_{i+1}} Y_n(s)\mu(s)ds$$
$$= E \int_0^\tau g(s)g'(s)Y_n(s)\mu(s)ds,$$

which is (11.9).

For a general g one would need to organize some sort of limiting process, along much the same lines as for Wiener integral in Lecture 10. Without going into it formally, we hope the reader will be more ready now to accept that the equation (11.9) will follow.

Lecture 12

Life insurance and net premiums

In this lecture we begin a new topic: life insurance policies and annuities. We will derive expressions for so called net premiums or pure premiums. All of our equations will be based on the distribution function F of the duration of life; therefore, in practical situations, the calculation of premiums considered below will use the estimates of F that we have been studying in previous lectures.

Let us begin with insurance.

Whole life insurance. In the simplest possible life insurance contract an individual A of age x purchases a life insurance policy, which guarantees that the beneficiaries of the policy will be paid a "sum insured" c in case of A's death. For this, person A assumes an obligation to pay the insurance company a constant premium p per annum until his death. How big should this premium be?

It is clear that A will pay only $p(T - x)$, which is, of course, a random variable. Calculating the expected value of this random variable and equating it to c we will obtain an expression for the average fair price:

$$p\,\mathsf{E}[T - x \,|\, T > x] = c,$$

i.e.

$$p_x = \frac{c}{\mathsf{E}[T - x \,|\, T > x]}. \tag{12.1}$$

The value of p, which equates the expected value of total premiums payable to the expected value of insurance payments receivable, is called the net premium; in non-life insurance the equivalent is often called the pure premium, or less frequently the risk premium, or occasionally the actuarially fair premium. To the net or pure premium in practice would be added loadings for expenses and uncertainties in the modelling of the risk covered, and possibly a profit margin. The final premium charged to the public is labelled the gross premium, usually specified as a premium rate, i.e. a premium rate per unit sum insured. See Neill [1989], ch. 2, for further discussion in a UK context; or Vaughan and Vaughan [2008], ch. 7, for a rather more detailed view in a North American context. We will restrict ourselves to considerations of the net premium and not discuss loadings superimposed on the net premium in practice.

If a person enters into a life insurance policy at the age of $x = 25$ for, say, $c = \$300000$, then $T - x$ is still quite large. For this person, therefore, it is important that because of inflation, \$300000 will not be nearly as large a sum in many years time, as it is at the moment when the contract is signed. To take this into consideration, let us introduce discounting, that is, let us assume that the sum \$a, which will be paid at one time unit later in the future, is equivalent to the sum $\$e^{-\rho} a$ today, with $\rho > 0$. We will also assume that payments are made continuously, as for water that flows into a pool. While it is not difficult to move from this abstraction to the reality of discrete payments, in continuous time everything looks clearer and more transparent. Besides, continuous payments are not as unrealistic in practice as one may think: banks calculate interest as a percentage on loans each day, and individuals often make fortnightly payments, i.e. quite frequently. So, let us set out the mathematical tools that we will be using.

The moment, or time 0 for us, is the time when the contract is signed and becomes valid.

The insured sum c, payable at the (random) moment $T - x$, in today's prices has value equal to $ce^{-\rho(T-x)}$, which has the expected value

$$c \mathsf{E}\left[e^{-\rho(T-x)} \,|\, T > x\right].$$

Here

$$\mathsf{E}\left[e^{-\rho(T-x)} \,|\, T > x\right] = \int_x^\infty e^{-\rho(t-x)} \frac{f(t)dt}{1 - F(x)}$$

$$= \int_0^\infty e^{-\rho s} \frac{f(s+x)}{1 - F(x)} ds \qquad (12.2)$$

(compare with (1.6)–(1.10)). The sum $p\,ds$, paid during the interval of time $[s, s+ds)$, in current prices is worth $e^{-\rho s} p\,ds$; and, more generally, in terms of money at time t it is worth $e^{\rho(t-s)} p\,ds$. Consequently, in current prices the value of the premium payable in the future is

$$p \int_0^{T-x} e^{-\rho s}\,ds = p \int_0^\infty e^{-\rho s} I_{\{T-x>s\}}\,ds. \qquad (12.3)$$

The expression under the right-hand side integral has a straightforward interpretation – during the interval of time ds, the premium paid is $p\,ds$ if an individual A is alive at the moment s, and 0 if A died before reaching s: i.e. the amount paid is $p\,ds\,I_{\{T-x>s\}}$. The expected value of (12.3) is

$$p\,\mathsf{E}\left[\int_0^{T-x} e^{-\rho s}\,ds \,\Big|\, T > x\right] = p \int_0^\infty e^{-\rho s}\,\mathsf{E}\left[I_{\{T-x>s\}} \,|\, T > x\right] ds$$

$$= p \int_0^\infty e^{-\rho s} \frac{1-F(x+s)}{1-F(x)}\,ds. \qquad (12.4)$$

Equating expected values (12.2) and (12.4),

$$p \int_0^\infty e^{-\rho s} \frac{1-F(x+s)}{1-F(x)}\,ds = c \int_0^\infty e^{-\rho s} \frac{f(x+s)}{1-F(x)}\,ds,$$

we again obtain the expression for the net premium:

$$p_{x,\rho} = c \frac{\int_0^\infty e^{-\rho s} f(x+s)\,ds}{\int_0^\infty e^{-\rho s}[1-F(x+s)]\,ds}. \qquad (12.5)$$

It is quite interesting to understand how the net premium depends on the discount rate ρ. In particular, which of the net premiums, (12.1) or (12.5), is the larger? Can we answer this question without knowing F? Essentially, yes, we can.

In particular,

if F is such that the corresponding force of mortality $\mu(y)$ increases in y for $y \geq x$, then the net premium (12.5) decreases in the discount rate ρ,

$$p_{x,\rho} > p_{x,\rho'}, \quad \text{if } \rho < \rho'; \qquad (12.6)$$

besides,

$$p_{x,\rho}|_{\rho=0} = \frac{c}{\mathsf{E}(T-x\,|\,T>x)}, \qquad p_{x,\rho}|_{\rho=\infty} = c\mu(x)\left(=c\frac{f(x)}{1-F(x)}\right).$$

The condition that $\mu(y)$ is an increasing function of y for $y > x$ is, of course, a very general condition. It is practically always true for the distribution functions used in demography (see, for example, Figure 2.1). While $\mu(y)$ may be decreasing for small values of y, and we may not be completely sure of its behaviour for y between, say, 10 and 25 years, for $y \geq 30$ the force of mortality $\mu(y)$ is increasing.

Thus, discounting decreases the net premium; and the higher the discount rate ρ, the greater the decrease.

Let us prove (12.6). Rewrite the integral in the right side of (12.5) in the following way

$$\int_0^\infty \frac{f(x+s)}{1-F(x+s)} \frac{e^{-\rho s}[1-F(x+s)]\,ds}{\int_0^\infty e^{-\rho y}[1-F(x+y)]\,dy}$$

$$= \int_0^\infty \mu(x+s)g_\rho(s)\,ds, \qquad (12.7)$$

where we set

$$g_\rho(s) = \frac{e^{-\rho s}[1-F(x+s)]}{\int_0^\infty e^{-\rho y}[1-F(x+y)]\,dy}.$$

But for any ρ, the function $g_\rho(s)$ is positive and its integral is equal to 1, so that $g_\rho(s)$ is the density of a distribution, dependent on ρ as a parameter. If $\rho < \rho'$, then

$$g_\rho(s) < g_{\rho'}(s) \quad \text{for } s < s_0,$$
$$g_\rho(s) > g_{\rho'}(s) \quad \text{for } s > s_0,$$

where s_0 is such that

$$g_\rho(s_0) = g_{\rho'}(s_0).$$

It is clear that

$$1 = \int_0^{s_0} g_{\rho'}(s)\,ds + \int_{s_0}^\infty g_{\rho'}(s)\,ds = \int_0^{s_0} g_\rho(s)\,ds + \int_{s_0}^\infty g_\rho(s)\,ds$$

and thus,

$$\int_0^{s_0} [g_{\rho'}(s) - g_\rho(s)]\,ds = \int_{s_0}^\infty [g_\rho(s) - g_{\rho'}(s)]\,ds.$$

Since, according to the assumption, $\mu(x+s)$ is increasing in s, we obtain

$$\int_0^{s_0} \mu(x+s)\big[g_{\rho'}(s) - g_\rho(s)\big]\,ds$$

$$\leq \mu(x+s_0)\int_0^{s_0}\big[g_{\rho'}(s) - g_\rho(s)\big]\,ds$$

$$= \mu(x+s_0)\int_{s_0}^\infty \big[g_\rho(s) - g_{\rho'}(s)\big]\,ds$$

$$\leq \int_{s_0}^\infty \mu(x+s)\big[g_\rho(s) - g_{\rho'}(s)\big]\,ds.$$

Rearranging this inequality, we obtain

$$\int_0^\infty \mu(x+s)g_{\rho'}(s)\,ds < \int_0^\infty \mu(x+s)g_\rho(s)\,ds.$$

And this, as you can see from (12.7), is indeed (12.6).

◇ **Exercise.** Obtain (12.4) without using equality (12.3), first proving the following integral

$$\int_0^{T-x} e^{-\rho s}\,ds = \frac{1}{\rho}\Big[1 - e^{-\rho(T-x)}\Big];$$

then taking the average with respect to the distribution function (1.8)

$$\frac{1}{\rho}\int_0^\infty (1 - e^{-\rho s})\,\frac{f(x+s)}{1-F(x)}\,ds;$$

and, finally, using integration by parts in the last integral. Prove that $p_{x,\rho} \to \mu(x)$ when $\rho \to \infty$. △

Whole life insurance with time-limited premiums. In contracts of this type a person A of age x still insures his life for an amount c; but a constant premium is payable only until a fixed time m, or until the moment of death $T - x$, should he die before time m. In other words, the premium is payable until the moment $\min(T - x, m)$. Overall, an amount paid as a premium by A, valuing at inception of the contract, equals

$$p\int_0^{(T-x)\wedge m} e^{-\rho s}\,ds = p\int_0^m e^{-\rho s} I_{\{T-x>s\}}\,ds, \qquad (12.8)$$

and its expected value is

$$p \int_0^m e^{-\rho s} \frac{1 - F(x+s)}{1 - F(x)} \, ds. \tag{12.9}$$

Therefore as an equation for the net premium we obtain

$$p \int_0^m e^{-\rho s} \frac{1 - F(x+s)}{1 - F(x)} \, ds = c \int_0^\infty e^{-\rho s} \frac{f(x+s)}{1 - F(x)} \, ds,$$

and hence

$$p_{x,\rho} = c \, \frac{\int_0^\infty e^{-\rho s} f(x+s) \, ds}{\int_0^m e^{-\rho s} [1 - F(x+s)] \, ds}. \tag{12.10}$$

Since the denominator now is smaller than in (12.5), the premium is now higher. This is of course as expected: the same benefit is being paid for over what is usually going to be a shorter period.

Term life insurance. In fixed term life insurance contracts the amount c is payable at the moment of death, if a person A of age x dies before the age of $x + m$. Otherwise, the insured sum is not payable and the premium is not refunded. Contracts of this type may be suitable for a person who will be placed in a more dangerous environment for a certain time. The value of such a contract in current prices at inception is equal to

$$ce^{-\rho(T-x)} I_{\{T-x<m\}},$$

and its expected value is

$$c \, \mathbb{E}\left\{ e^{-\rho(T-x)} I_{\{T-x<m\}} \,\middle|\, T > x \right\} = c \int_0^m e^{-\rho s} \frac{f(s+x)}{1 - F(x)} \, ds. \tag{12.11}$$

The value of the overall premium paid in this contract is the same as in the previous case, and equals (12.8) with expectation (12.9). Equating the values of income and benefits we obtain the equation for the net premium

$$p \int_0^m e^{-\rho s} [1 - F(x+s)] \, ds = c \int_0^m e^{-\rho s} f(x+s) \, ds. \tag{12.12}$$

It is obvious how this equation can be generalized to the situation in which person A is paying a premium until the time $\min(T - x, m_1)$, while the sum insured is payable only if A does not live until the age $x + m_2$, where $m_2 > m_1$. In this case in the integral on the left m should be replaced by m_1; and in the integral on the right it should be replaced by m_2:

$$p \int_0^{m_1} e^{-\rho s}[1 - F(x+s)]\,ds = c \int_0^{m_2} e^{-\rho s} f(x+s)\,ds. \qquad (12.13)$$

If $m_2 \to \infty$ or $m_1, m_2 \to \infty$ this equation is transformed into equations (12.10) and (12.5), respectively.

It is interesting to let $m \to 0$ in equation (12.12), i.e. when insurance is taken for a very short term. Since integrated functions behave smoothly, equation (12.12) leads to

$$p = c\frac{f(x)}{1 - F(x)}, \qquad (12.14)$$

so that the premium becomes proportional to the force of mortality. This "short term" insurance is usually a one-year contract. If the premium is paid at the beginning of the year, while the insured amount is paid, if at all, at the end of the year, then equation (12.12) has to be replaced by

$$p[1 - F(x)] = ce^{-\rho}\Delta F(x),$$

i.e.

$$p = ce^{-\rho}\frac{\Delta F(x)}{1 - F(x)}, \qquad (12.15)$$

which needs to be adjusted every year.

◇ **Exercise.** Figure 12.1 shows term insurance premiums for a life year of a leading Australian life insurance company at different ages, for the year 1999; also shown are the values of the force of mortality $\Delta F(x)/[1 - F(x)]$, with the step of one year, for the population of Australia in the same year, as obtained from the Australian Bureau of Statistics. As we see the premium is systematically smaller than the force of mortality for all ages. Why would the company request a lower premium than it apparently should? See formula (12.15) and the remark on mixtures at the end of the lecture. △

We have an example of insurance with a variable premium when a person A renews his insurance every year.

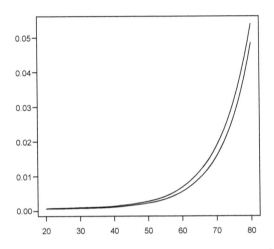

Figure 12.1 *National yearly force of mortality (solid line) and gross premiums (per $1 sum insured) for insurance for one year at ages 20 to 80 years.*

Consider again a fixed interval $[0,m]$ of a term life insurance contract and partition it into subintervals $[t_i, t_{i+1})$, $i = 0, 1, \ldots, N-1$, of, say, equal lengths $\Delta = t_{i+1} - t_i = m/N$, so that $t_i = i\Delta$. Suppose that the premium is paid at the beginning of each interval, while the insured sum is payable, if at all, at the end of each period. In this case for an interval $[t_i, t_{i+1})$ the premium $p\Delta$ will be paid with probability $[1 - F(x+t_i)]/[1 - F(x)]$ and should be discounted by the factor $e^{-\rho t_i}$; while the amount insured will be paid with probability $\Delta F(t_i)/[1 - F(x)]$ and should be discounted by the factor of $e^{-\rho t_{i+1}}$. We see that, for each interval,

$$\int_{t_i}^{t_{i+1}} e^{-\rho s}[1 - F(x+s)]\, ds < e^{-\rho t_i}[1 - F(x+t_i)]\Delta,$$

and

$$\int_{t_i}^{t_{i+1}} e^{-\rho s} f(x+s)\, ds > e^{-\rho t_{i+1}} \Delta F(x+t_i),$$

so that we have

$$\int_0^m e^{-\rho s}[1 - F(x+s)]\,ds < \sum_{i=0}^{N-1} e^{-\rho t_i}[1 - F(x+t_i)]\Delta,$$

$$\int_0^m e^{-\rho s}f(x+s)\,ds > \sum_{i=0}^{N-1} e^{-\rho t_{i+1}}\Delta F(x+t_i). \tag{12.16}$$

The net premium for such discrete payments is now given by the equation

$$p\sum_{i=0}^{N-1} e^{-\rho t_i}[1 - F(x+t_i)]\Delta = c\sum_{i=0}^{N-1} e^{-\rho t_{i+1}}\Delta F(x+t_i),$$

and has become somewhat smaller than the premium in the case of continuous payments defined by (12.13). This is of course to be expected: we have advanced the premium payments, and delayed the payments of the benefit.

Insurance with variable sum insured and variable premium.
The previous insurance contracts considered, with fixed premium and fixed sums insured, are called "level cover" contracts. Consider now the case when, say, the amount payable in the event of death is not fixed, but changes over time in a prescribed way.

Usually the sum insured increases over time, increasing each year by a "bonus" under so-called "with-profit" life insurance contracts. The calculation of the bonus proceeds either arithmetically (a simple reversionary bonus) or geometrically (a compound reversionary bonus).

Under the former type of bonus, and tacitly disregarding the fact that such bonuses are normally granted to the policy only annually, the sum insured $c(t)$ increases linearly through time:

$$c(t) = c + bt.$$

The amount $c(T-x)$, payable at the time of death, discounting to the inception of the insurance policy, has value

$$c(T-x)e^{-\rho(T-x)} \tag{12.17}$$

and its expected value is

$$E\left[c(T-x)e^{-\rho(T-x)}\,|\,T>x\right] = \int_0^\infty c(s)e^{-\rho s}\frac{f(x+s)\,ds}{1-F(x)}. \qquad (12.18)$$

For example, for the linear $c(t)$ the expected value (12.18) can be written as

$$c\int_0^\infty e^{-\rho s}\frac{f(x+s)}{1-F(x)}\,ds + b E\left[(T-x)e^{-\rho(T-x)}\,|\,T>x\right].$$

Analogously, when the premium is changing over time in a prescribed way, that is, it becomes a given function $p(t)$ of time, the value at policy inception of the overall premiums paid is

$$\int_0^\infty p(s)e^{-\rho s}I_{\{s<T-x\}}\,ds$$

with expected value

$$\int_0^\infty p(s)e^{-\rho s}\frac{1-F(x+s)}{1-F(x)}\,ds. \qquad (12.19)$$

One can treat various particular forms of $c(t)$ and $p(t)$, but the most important case is when both $c(t)$ and $p(t)$ grow continuously in time as

$$c(t) = ce^{\rho_1 t} \quad \text{and} \quad p(t) = pe^{\rho_2 t}.$$

That is, the sum insured payable increases by a compound reversionary bonus operating continuously at rate ρ_1, while the premium also increases continuously at rate ρ_2; and the rates ρ_1 and ρ_2 do not have to be equal.

Rate ρ_2 is typically related to, or simply equal to, the so-called consumer price index (CPI) or inflationary index; while ρ_1 may, or may not, include a pure bonus component in addition to the inflationary expectations. The most natural case is when these rates are equal: $\rho_1 = \rho_2 = \rho$.

In this case both expected values in (12.18) and (12.19) turn out to be $p E[T-x\,|\,T>x]$ and c, and equating them we get exactly the same relationship between p and c as in the whole of life insurance contract considered at the beginning of this lecture. However, now we can apply the proposition (12.6) and see that in the case of compounding $c(t)$ and $p(t)$ the ratio p/c is larger than in the case of level cover with constant c and p. This does not imply, of course, that the contract in the compounding case is "more expensive" – in both contracts the expected values of what the policyholder and insurance company pay are equal; but in the case of the compounded amounts larger sums are changing hands.

Group life insurance. Consider a group of k persons of ages $x_1, x_2, \ldots, x_k$ entering into a joint life insurance contract with an insurance company. Let the distribution functions of the remaining lifetimes $T_1 - x_1, T_2 - x_2, \ldots, T_k - x_k$ be $F_{x_1}, F_{x_2}, \ldots, F_{x_k}$. That is,

$$1 - F_{x_1}(s) = P\{T_1 - x_1 > s \,|\, T_1 > x_1\},$$

and similarly for the other members of the group. Individuals of this group, for example, members of a tourist group on the same long journey, will have different genders, will be in different states of health, etc. So, although we have

$$1 - F_{x_1}(s) = \frac{1 - F_1(x_1 + s)}{1 - F_1(x_1)},$$

$$1 - F_{x_2}(s) = \frac{1 - F_2(x_2 + s)}{1 - F_2(x_2)},$$

and so on, we do not suppose that the distribution functions $F_1, F_2, \ldots, F_k$ are all identical. Let us agree, that according to the contract, the sum insured is payable at the time of the first death in this group. This means, of course, that the sum insured is paid at the random moment

$$\tau = \min(T_1 - x_1, \ldots, T_k - x_k).$$

If random variables $T_1 - x_1, T_2 - x_2, \ldots, T_k - x_k$ are independent, then

$$1 - G(s) = P\{\tau > s\} = [1 - F_{x_1}(s)] \cdots [1 - F_{x_k}(s)].$$

While it is possible to have different contracts with different times of payments of the sum insured, analysis of the large number of possible such contracts is not our purpose here. What is important for us though, is to note the following principle.

Once we have defined when the sum insured is payable and found the distribution function $G(s)$ of the time of payment, all of our previous work remains valid upon the substitution of the conditional distribution function $F_x(s) = [F(x+s) - F(x)]/[1 - F(x)]$ by the function $G(s)$. For example, in the case of a fixed term insurance the equation for the net premium looks very similar to that in equation (12.12):

$$p \int_0^m e^{-\rho s}[1 - G(s)] \, ds = c \int_0^m e^{-\rho s} g(s) \, ds.$$

It is just that $1 - G(s)$ has replaced $[1 - F(x+s)]/[1 - F(x)]$ and the density $g(s)$ is no longer $f(x+s)/[1 - F(x)]$.

◇ **Exercise.** From (12.1), write down an equation for the net premium without discounting for a homogeneous group when all of the F_{x_i} are equal; and the payment of premium stops, and the sum insured c is payable, upon the first death.   △

We conclude the list of examples of insurance contracts with two remarks.

For a level cover contract, the net premium p is a simple multiple of the sum insured c. In the outworkings of this case one needs, therefore, to consider only the net premium per dollar insured, or p/c. We think, however, that working with c and p separately enhances one's understanding of the situation. In any case, the simple proportionality of premium and sum insured no longer necessarily obtains when the gross premium is considered; moreover, in more complicated situations in which the sum insured and/or the premium vary over time, one would need to consider c and p separately.

The expected value of the contract at its inception is called the actuarial value of the contract. In particular, (12.2) is an actuarial value of the contract (with $c = 1$); and the right-hand side of (12.18) also represents the actuarial value of the corresponding contract.

12.1* Remark on mixtures of distributions

The payment upfront for periodic cover in (12.15) leads to lower premiums. The effect, however, is not that strong.

The main factor in the rather paradoxical situation, when an insurance company charges a gross premium smaller than what should have been a net premium for a general population, is something different. This factor is that a socio-economic group of individuals, interested in life insurance and able to afford it, differs in habits and lifestyle from the general population. This phenomenon is referred to as "selection" in the insurance industry (see, e.g., Neill [1989]).

On the one hand, this explains why insurance companies keep and analyze their own statistical records, rather than relying on national data – this was what enabled the company in Figure 12.1 to offer competitive prices to the customers. On the other hand, this confirms the point of view of a general population as a mixture of different layers and socio-economic groups with possibly quite different distributions of lifetimes.

Mixtures occur in the study of lifetimes (or failure times) quite often. For example, speaking about failure times of a device, a small change in production technology, which may remain unknown to a user of the device, could easily change the distribution of its failure times. Hence, the user is dealing with a mixture of devices made according to one or the other technology. Sometimes it can also be the change in the manufacturer, rather than in technology.

In demography mixtures occur very frequently. For example, observations of lifetimes of individuals who died in a given year is actually an observation on a mixture: these individuals were born in different years and therefore, in general, belong to different cohorts. Are there really differences in cohorts which are close in time, that is, is the population stationary? This theme leads to interesting mathematical problems.

One more example when we deal with a mixture is connected with Figure 6.2, which shows graphs of the survival functions of the male population of New Zealand in 1876, 1900 and 1932.

The population of New Zealand can be subdivided into at least two groups: descendants of Maori, who began colonization of both the North and South Islands between about the 8th and the 12th centuries; and descendants of European settlers, Pakeha, who began to settle in New Zealand more or less systematically from the end of the 18th and the beginning of the 19th centuries. The boundary between Maori and Pakeha is blurred by a large number of mixed marriages, but in the years of which we speak the distinction was more pronounced.

It is natural to assume that the distribution functions for the male populations of Maori, F_1, and Pakeha, F_2, are not the same. Then overall observations on the national mortality will actually be observations on random variables with the distribution function

$$G(x) = pF_1(x) + (1-p)F_2(x),$$

which is a mixture of F_1 and F_2; and in which p denotes the fraction of the male population of Maori.

If, however, this is true, then the changes in G, which are so clearly demonstrated in Figure 6.2, may arise for two distinct reasons: either a change in F_1 and F_2, when lifetimes in both groups became longer; or a change in the proportion p only. Merely by looking at Figure 6.2 one cannot be sure that an apparent improvement in national longevity is not simply a consequence of an increase in the proportion of the longer-lived of the groups. Additional investigation is necessary.

In this case, the additional investigations have indeed been made, asserting that the fraction p of the Maori population decreased, mostly due to European migration, from 12% in 1876 to around 5.5% in 1901, decreased a little bit more by 1921 to 4.5% and in 1936 again reached 5.2%. The data vary slightly between different sources, but the overall picture stays the same, see, e.g., Dunstan et al. [2006], Jordan [1921], Neale [1940]; also the difference between F_1 and F_2 exists, both of these distribution functions changed, and durations of life in both groups essentially increased. In particular, child mortality among Maori decreased.

Consider a simple mathematical model of a mixture. It will very well illustrate the new phenomenon that occurs when we consider populations as a mixture.

Suppose $\mu(x)$, $x > 0$, is a "baseline" mortality rate and suppose that each lifetime of each member of a population has the mortality rate $Z\mu(x)$ with its own, specific to an individual, value of $Z > 0$. That is, the duration of life of each individual follows the distribution function with the tail

$$1 - F_Z(x) = e^{-Z \int_0^x \mu(y)dy}.$$

When the value of Z becomes large, $1 - F_Z(x)$ becomes small at every x and F_Z is more concentrated near 0. Hence, the corresponding lifetime becomes smaller, or shorter, in probability. When the value of Z becomes small the opposite effect occurs: $1 - F_Z(x)$ becomes close to 1 at every x, and F_Z is more spread over the entire positive half-line, so that the corresponding lifetimes become larger in probability. Suppose the values of Z are spread in a population according to some distribution function $H(z)$, and assume this $H(z)$ has a density $h(z)$, which is finite, positive and continuous at $x = 0$. The latter is a purely technical assumption and will help to keep the derivations simple. The H itself is in most cases unknown, even if we are certain that we deal with a mixture, but the conclusion we will arrive at will not depend on particular forms of H.

A randomly selected individual A will have a randomly selected Z. If we knew the value of this Z, or if we wanted to condition on this Z, the distribution function of the lifetime of A would be F_Z. Since we do not know it and speak about a randomly selected individual, the distribution function of A's lifetime becomes a mixture

$$1 - F(x) = \int_0^\infty [1 - F_z(x)]dH(z) = \int_0^\infty e^{-z\int_0^x \mu(y)dy}dH(z).$$

If we have a sample of lifetimes $T_1, \ldots, T_n$ from this mixed population, we have a sample from F. The force of mortality, which this sample will reveal, is the force of mortality, corresponding to F, which is

$$\eta(x) = -\frac{d}{dx}\ln[1 - F(x)],$$

and which certainly does not have the form $z\mu(x)$. Consider the asymptotic behavior of $\eta(x)$ for large ages x. What we obtain is this statement:

for the force of mortality of the mixture F above the following asymptotic expression is true:

$$\eta(x) \sim \frac{\mu(x)}{\int_0^x \mu(y)dy}, \quad \text{as} \quad x \to \infty. \tag{12.20}$$

Recall that $\int_0^x \mu(y)dy \to \infty$, as $x \to \infty$, for any force of mortality (see Lecture 1). Therefore, an immediate corollary is that

$$\eta(x) = o(\mu(x)), \quad \text{as} \quad x \to \infty.$$

That is, in a mixture, people would seem to live longer than it could be expected from each individual. This effect is very strong. If, for example, $\mu(x)$ is the failure rate of the Weibull distribution, see (2.2), which is increasing with age as x^{k-1}, for η we obtain

$$\eta(x) \sim \frac{k}{x}, \quad \text{as} \quad x \to \infty.$$

Hence, η is decreasing, and the power of x is the same for any power k in the Weibull distribution. If, as another example, $\mu(x)$ is the failure rate of the Gompertz distribution, see (2.4), which increases with age as an exponent e^{cx}, for η we obtain

$$\eta(x) \sim \frac{\theta c e^{cx}}{\theta(e^{cx} - 1)} \sim c, \quad \text{as} \quad x \to \infty.$$

Therefore, the tail of F will behave as the tail of exponential distribution function, i.e. totally differently from (2.5).

Now we show the proof of (12.20). Let $\Lambda(x) = \int_0^x \mu(y)dy$ and note that the density of F has the form

$$f(x) = \mu(x) \int_0^\infty z e^{-z\Lambda(x)} dH(z).$$

Choose small $\varepsilon > 0$ and split the integral into two:

$$\int_0^\varepsilon z e^{-z\Lambda(x)} dH(z) + \int_\varepsilon^\infty z e^{-z\Lambda(x)} dH(z).$$

For sufficiently large x, such that $\varepsilon > 1/\Lambda(x)$, the function $z e^{-z\Lambda(x)}$ will be decreasing on $[\varepsilon, \infty)$ and therefore

$$\int_\varepsilon^\infty z e^{-z\Lambda(x)} dH(z) \le \varepsilon e^{-\varepsilon\Lambda(x)}.$$

In the first integral we change the variable $t = z\Lambda(x)$ and use the assumption on the density $h(x)$:

$$\int_0^\varepsilon z e^{-z\Lambda(x)} h(z)dz = \frac{1}{\Lambda^2(x)} \int_0^{\varepsilon\Lambda(x)} t e^{-t} h\left(\frac{t}{\Lambda(x)}\right) dt$$

$$\sim \frac{1}{\Lambda^2(x)} h(0) \int_0^{\varepsilon\Lambda(x)} te^{-t} dt.$$

With $\varepsilon\Lambda(x) \to \infty$ the integral converges to 1. Since $e^{-\varepsilon\Lambda(x)}$ is infinitely smaller than $1/\Lambda(x)^2$, the second integral is negligible, and eventually we obtain:

$$f(x) \sim \mu(x) h(0) \frac{1}{\Lambda^2(x)}, \quad \text{as} \quad x \to \infty.$$

In a very similar way we can show that

$$1 - F(x) \sim h(0) \frac{1}{\Lambda(x)}, \quad \text{as} \quad x \to \infty.$$

Therefore

$$\frac{f(x)}{1 - F(x)} \sim \frac{\mu(x)}{\Lambda(x)},$$

which is (12.20).

◇ **Exercise.** a) Derive the asymptotic representation for $1 - F(x)$ above.

b) In populations with $h(0) > 0$ and $h(0) = 0$ the proportion of people who would live a long time is different and in the latter it is smaller. Assume that in ε-neighborhood of zero $h(z) \sim z^\beta$ with any positive β and show that our statement (12.20) will, basically, not change:

$$\frac{f(x)}{1 - F(x)} \sim (\beta + 1) \frac{\mu(x)}{\Lambda(x)},$$

△

Lecture 13

More on net premiums. Endowments and annuities

In this lecture we will discuss further types of insurance contracts, particularly annuities, for which we will derive net premiums. We would like, however, to caution against the impression that the calculation of net premiums is logically as clear and straightforward as it has so far been presented.

In reality, the sum insured is paid at a random moment τ: in Lecture 10 this random moment was either $T - x$ for an individual person of age x at inception of the policy, or had a somewhat more complicated structure for a group insurance. The amount paid in prices in force at the inception of the contract is equal to $ce^{-\rho\tau}$, with an expected value of

$$\pi_0 = c\,\mathsf{E}e^{-\rho\tau}.$$

This is called the actuarial value of the insurance policy, equal to a "one-off" net premium if one is paying for the life insurance contract entirely "up-front": such a single premium, especially in the purchase of annuities, is often called a "consideration" (see Vaughan and Vaughan [2008], p. 167, for legal aspects of this term). Returning to the mathematics, however, the value of this premium π_0 at the moment of payment of the sum insured c will be

$$\pi_0 e^{\rho\tau}$$

and, consequently, its expected value will be

$$\pi_0\,\mathsf{E}e^{\rho\tau} = c\,\mathsf{E}e^{-\rho\tau}\,\mathsf{E}e^{\rho\tau},$$

which is more than c! Indeed, according to the Cauchy–Bunyakovsky inequality

$$1 = \mathsf{E}e^{-\frac{\rho\tau}{2}}e^{\frac{\rho\tau}{2}} < \mathsf{E}e^{-\rho\tau}\mathsf{E}e^{\rho\tau}, \tag{13.1}$$

so that

$$\pi_0 \mathsf{E}e^{\rho\tau} > c. \tag{13.2}$$

Thus, what appears to be the net premium at price levels current at inception, becomes an overpayment when expressed at price levels in force at termination or maturity of the contract. Note that

$$\pi_\tau = \frac{c}{\mathsf{E}e^{\rho\tau}}$$

is the net premium in prices in force at the moment τ, i.e. at termination. The same is true when premiums are paid during the whole of life, or during a limited period. Indeed, in the first case the overall sum paid expressed in prices current at the moment τ is (see (12.3))

$$pe^{\rho\tau}\int_0^\tau e^{-\rho s}\,ds$$

and its expected value is

$$p\,\mathsf{E}e^{\rho\tau}\int_0^\tau e^{-\rho s}\,ds.$$

But, because

$$\mathsf{E}e^{\rho\tau}\int_0^\tau e^{-\rho s}\,ds > \mathsf{E}e^{\rho\tau}\mathsf{E}\int_0^\tau e^{-\rho s}\,ds, \tag{13.3}$$

the premium $p_{x,\rho}$, defined by (12.13), is again too high:

$$c\,\frac{\mathsf{E}e^{-\rho\tau}}{\mathsf{E}\int_0^\tau e^{-\rho s}\,ds}\cdot\mathsf{E}e^{\rho\tau}\int_0^\tau e^{-\rho s}\,ds > c\,\mathsf{E}e^{-\rho\tau}\mathsf{E}e^{\rho\tau} > c. \tag{13.4}$$

◇ **Exercise.** a) What is the expression for the net premium when we equate expected values at the moment of payment?

b) Prove the inequality (13.3) by taking the integrals and using (13.1). △

In insurance mathematics we note that there is a whole theory devoted to the definition of net premiums based on a general loss function or utility function $u(\pi - X)$, which depends on a single 'one-off' premium π and stochastic repayments X. Namely, a unique net premium is defined as the root of the equation

$$Eu(\pi - X) = 0 \qquad (13.5a)$$

or as the root of the more general equation

$$Eu(\pi, X) = 0 \qquad (13.5b)$$

(see, e.g., Goovaerts et al. [1984] and also Denuit et al. [1999]); and each new choice of loss function leads to a new value of the net premium. In mathematical statistics equations (13.5) and their empirical analogues are used to derive so-called M-estimators (see, for example, Huber [1981]). We note, however, that the problem raised above by inequalities (13.2) and (13.4) is not connected with the choice of loss function. It is rather connected with the existence of different values of the same contract, and with the fact that discounting from a random moment and averaging are not permutable or interchangeable operations. Phenomena of this type, see (13.2) and (13.4), do not occur for a deterministic moment t: the equalities

$$pe^{\rho t} \int_0^t e^{-\rho s}\, ds = c, \qquad p \int_0^t e^{-\rho s}\, ds = ce^{-\rho t}$$

are equivalent.

Let us now consider annuities and calculate the corresponding net premiums or considerations at the start of the contract. After all, a contract is entered into using the price levels at its inception.

Pure endowment. In this contract a person A of an age x will receive the amount c if the person lives until the age $x + m$. Otherwise the amount is not payable and the premium is not returned. In other words, person A receives the amount

$$cI_{\{T-x\geq m\}}$$

at the moment m, as distinct from the term life insurance contract, where the amount $cI_{\{T-x<m\}}$ is payable at the moment $T - x$. The expected value of the future payment under the (pure) endowment, expressed in the price level at inception, is

$$ce^{-\rho m} E\left[I_{\{T-x>m\}} \mid T > x\right] = ce^{-\rho m} \frac{1 - F(x+m)}{1 - F(x)}, \qquad (13.6)$$

which is the actuarial value of this contract. If, for this payment, the person is paying the premium up to the moment m, the person A will pay on average, valuing at the initial price level,

$$p E\left[\int_0^m e^{-\rho s} I_{\{T-x>s\}} \, ds \mid T > x\right] = p \int_0^m e^{-\rho s} \frac{1 - F(x+s)}{1 - F(x)} \, ds, \qquad (13.7)$$

which is exactly the expression given in (12.9). Equating (13.6) and (13.7), one obtains an equation for the net premium:

$$p \int_0^m e^{-\rho s}[1 - F(x+s)] \, ds = ce^{-\rho m}[1 - F(x+m)]. \qquad (13.8)$$

This equation leads to values of p different from those obtained from (12.12). This is illustrated in the following exercise.

◇ **Exercise.** If the person A, who reaches the age $x+m$, was simply to deposit the amount of premiums (13.8) in a bank, starting at inception and paying continuously, as above, with the same interest rate ρ, he would obtain only the amount

$$c \frac{1 - e^{-\rho m}}{\rho} \frac{1 - F(x+m)}{\int_0^m e^{-\rho s}[1 - F(x+s)] \, ds},$$

which is smaller than c, the amount the person will get under the pure endowment contract. Verify this.   △

◇ **Exercise.** a) What would be the form of the equation (13.8) at the price level at the moment m?

b) A man of age 55 years is buying a pure endowment for maturity in 5 years time. According to mortality tables for the male population of New Zealand the annual mortality rates

$$\frac{\Delta \widehat{F}(y)}{1 - \widehat{F}(y)}$$

over 2005-2007 for ages $y = 55, 56, \ldots, 59$ years have been

$$0.00501, \ 0.00549, \ 0.00602, \ 0.00660, \ 0.00752,$$

while

$$1 - \widehat{F}(55) = 0.92606.$$

Assuming that the premium is paid at the beginning of each year of the contract, derive the discrete analogue of the equations (12.12) and (13.8), and find numerical values of p per dollar (or assuming $c = \$100,000$). Choose $\rho = 0.03$, slightly higher than the current inflation rate in new Zealand. $\triangle$

Life annuity. In this case person A periodically (yearly, monthly, etc.) receives a regular payment of R. So, ignoring any discounting, person A receives the sum

$$R(T - x),$$

and the value of this sum at the initial price level is equal to

$$R \int_0^{T-x} e^{-\rho s} \, ds = R \int_0^{\infty} e^{-\rho s} I_{\{T-x>s\}} \, ds.$$

Its expected value, i.e. its actuarial value, is equal to

$$R \int_0^{\infty} e^{-\rho s} \frac{1 - F(x+s)}{1 - F(x)} \, ds, \tag{13.9}$$

which coincides with the expression (12.4). In order to illustrate the terminology, rather than anything else, let us consider the discrete analogue of the integral (13.9): let us break the time half-line $[0, \infty)$ into intervals $[t_i, t_{i+1})$, $i = 0, 1, \ldots, n$, $t_0 = 0$. If payments (rents) are paid at the beginning of each interval, the annuity is called an annuity-due [Neill, 1989, p. 41]. The actuarial value of an annuity-due is

$$\pi_R = R \sum_{i=0}^{\infty} e^{-\rho t_i} \frac{[1 - F(x+t_i)]}{1 - F(x)} \Delta t_i,$$

which is more than the integral (13.9). If payments are made at the end of each interval, the annuity is called an ordinary annuity, and its actuarial value is

$$\pi_R' = R \sum_{i=0}^{\infty} e^{-\rho t_{i+1}} \frac{[1 - F(x + t_{i+1})]}{1 - F(x)} \Delta t_{i+1},$$

which is less than the integral in (13.9). It is clear that

$$\pi_R = \pi_R' + R\Delta t_0.$$

Deferred annuity. When there is no delay in receipt of the annuity payments, the annuity is called an immediate annuity (see, e.g., Neill [1989], p. 40, and Vaughan and Vaughan [2008], p. 320); otherwise we refer to the annuity as a deferred annuity. In this case the same regular payments are made as above, but not immediately after buying an annuity: payments commence after a certain time l. Thus, person A will nominally receive, in the absence of discounting,

$$R(T - x - l)^+ = \begin{cases} R(T - x - l), & \text{if } T - x > l; \\ 0, & \text{if } T - x \leq l. \end{cases}$$

It is clear that the actuarial value of the deferred annuity is

$$RE\left[\int_0^{\infty} e^{-\rho(s+l)} I_{\{T - x > l + s\}} \, ds \, \middle| \, T > x \right]$$

$$= Re^{-\rho l} \int_0^{\infty} e^{-\rho s} \frac{1 - F(x + l + s)}{1 - F(x)} \, ds.$$

This can be rewritten in the following more convenient form:

$$e^{-\rho l} \frac{1 - F(x + l)}{1 - F(x)} \cdot R \int_0^{\infty} e^{-\rho s} \frac{1 - F(x + l + s)}{1 - F(x + l)} \, ds,$$

which emphasizes that the actuarial value of the annuity (at the initial price level) is equal to its actuarial value at the moment l, times the probability of survival to the age $x + l$, and discounted by $e^{-\rho l}$.

Term annuity. In contrast to a life annuity, payments on a term annuity are made only until the age $x+m$. This will be covered as a special case of the annuity described next.

Term deferred annuity. This is the most general of all the cases considered so far. Payments of this annuity are periodic (yearly, monthly, etc.) beginning from the age $x+l$ until the age $x+m$, save that any payment ceases upon death; and indeed should the individual die before age $x+l$, nothing is paid at all. The amount paid in the absence of discounting will be equal to

$$R \int_l^m I_{\{T-x>s\}} \, ds,$$

and its value at the initial price level is equal to

$$R \int_l^m e^{-\rho s} I_{\{T-x>s\}} \, ds.$$

Therefore, the actuarial value of the term deferred annuity is equal to

$$R \int_l^m e^{-\rho s} \frac{1-F(x+s)}{1-F(x)} \, ds.$$

For $l = 0$ this formula gives us the actuarial value of the "ordinary" m-year term annuity, while for $l = 0$ and $m = \infty$ we obtain the actuarial value of the whole-life annuity.

◇ **Exercise.** As for the case of group insurance (Lecture 10), consider the possibilities for annuities for a group: these are the joint-life annuity, the last survivor annuity, etc. △

Endowment assurance contracts. In contracts of this type an amount c_1 is paid to the policy beneficiaries of an insured person, if the latter dies before age $x+m$; and a sum insured (or sum assured) c_2 is paid to the insured person should that person survive until the age of $x+m$. In the traditional endowment assurance contract the sums c_1 and c_2 were equal (see, e.g., Neill [1989], p. 49). The amount payable without any discounting equals

$$c_1 I_{\{T-x<m\}} + c_2 I_{\{T-x \geq m\}}.$$

The value of this amount at the price level at inception equals

$$c_1 e^{-\rho(T-x)} I_{\{T-x<m\}} + c_2 e^{-\rho m} I_{\{T-x\geq m\}},$$

so that its actuarial value is the sum of the actuarial values of a term life insurance contract (see (12.11)) and a pure endowment (see (13.6))

$$c_1 \mathsf{E}\left[e^{-\rho(T-x)} I_{\{T-x<m\}} \,\middle|\, T > x\right] + c_2 e^{-\rho m} \mathsf{E}\left[I_{\{T-x\geq m\}} \,\middle|\, T > x\right]$$

$$= c_1 \int_0^m e^{-\rho s} \frac{f(x+s)}{1-F(x)}\,ds + c_2 e^{-\rho m} \frac{1-F(x+m)}{1-F(x)}.$$

In the same way a single net premium, or a net premium payable for a fixed term, for this "mixed" contract will be equal to the sum of the analogous net premiums for each of the contracts.

Lecture 14

Annuities certain. Some problems of general theory

It seemed to us worthwhile to spend two lectures on particular forms of insurance contracts, because we now wish to try to see the wood for the trees by looking at the underlying principles common to all contracts considered so far. From a rather crowded literature one can mention Gerber [1979], p. 39, who treats profits emerging over time from a life insurance contract in a probabilistic setting; Neill [1989], ch. 4, which contains a comprehensive account of life insurance policy values or policy reserves and how their values move over time, from a traditional actuarial viewpoint, which essentially treats the empirical distribution function as the population distribution function; and Vaughan and Vaughan [2008], ch. 8, in which policy reserves in life insurance are put in the context of unearned premium reserves in general insurance, and the discussion takes place firmly in a North American context.

Let us start with the following expression: for a random time τ with a distribution function G consider

$$cI_{\{\tau<t\}} - \int_0^t p(s)I_{\{\tau>s\}}\,ds. \tag{14.1}$$

This is the difference between undiscounted amounts of c payable at time τ by an insurance company and what was received by the company in continuously paid premiums $p(s)$ up to the moment $\min(t,\tau)$. It gives us the loss, or gain if it is negative, of the company as at the moment of time t. If so, then

$$e^{-\rho s}[c\,dI_{\{\tau<s\}} - p(s)I_{\{\tau\geq s\}}\,ds] \tag{14.2}$$

167

is the value of a loss during the time interval ds expressed in prices at inception, while

$$c \int_0^t e^{-\rho s} \left[dI_{\{\tau < s\}} - \frac{p(s)}{c} I_{\{\tau \geq s\}} ds \right] \qquad (14.3)$$

is the accumulated value for all losses up to time t, again expressed at the initial price level. The value of these losses at the price level currently in force at the moment t is $e^{\rho t}$ times larger than (14.3) and equals

$$c \int_0^t e^{\rho(t-s)} \left[dI_{\{\tau < s\}} - \frac{p(s)}{c} I_{\{\tau > s\}} ds \right]. \qquad (14.4)$$

All equations for net premiums in Lectures 12 and 13 were derived from the condition that the expected value of (14.3) must be zero.

If there are n independent contracts of the same type, where the random moments of payments $\tau_1, \tau_2, \ldots, \tau_n$ are independent and identically distributed (with the same distribution function G), then summing equations (14.1), (14.2) and (14.4) over different contracts, that is, over $i = 1, 2, \ldots, n$, we obtain the undiscounted losses of the portfolio:

$$c z_n(t) - \int_0^t p(s)[n - z_n(s)] ds, \qquad (14.5)$$

while its value in initial prices and current prices (at time t) are

$$c \int_0^t e^{-\rho s} \left[dz_n(s) - \frac{p(s)}{c} (n - z_n(s)) ds \right] \qquad (14.6a)$$

and

$$c \int_0^t e^{\rho(t-s)} \left[dz_n(s) - \frac{p(s)}{c} (n - z_n(s)) ds \right], \qquad (14.6b)$$

respectively. Here

$$z_n(t) = \sum_{i=1}^n I_{\{\tau_i < t\}}$$

is a binomial process, familiar to us from Lectures 3–5. In particular, as we know – see b) in the penultimate exercise in Lecture 9 – the process

$$dm_n(s) = dz_n(s) - \frac{g(s)}{1 - G(s)} [n - z_n(s)] ds$$

is a martingale. In other words, if we require that the expected value of (14.6b) be zero for all t, then this will imply that $p(s) = cg(s)/[1 - G(s)]$. This is exactly the net premium of a short-term life insurance. Hence one can say that if the life insurance is arranged through a sequence of many short-term contracts with premium per dollar $g(s)/[1 - G(s)]$, then the undiscounted losses and the losses at the initial price level are martingales. If the premium $p(s)/c$ were chosen differently, then these processes in t would no longer be martingales — they would acquire a drift.

The process (14.6b) is however not a martingale: it has a different structure. Consider it more closely.

Let us forget for a while that (14.6) is a transformation of a random process. Let us assume, instead, that some function $a(t)$ defines certain payments, so that the amount $a(s)\,ds$ is paid during the time interval $[s, s+ds)$. Then the value of all payments made, evaluated at the price level in force at inception, is

$$\int_0^t e^{-\rho s} a(s)\,ds,$$

while in the prices of the current moment t it is

$$\varphi(t) = \int_0^t e^{\rho(t-s)} a(s)\,ds.$$

This function $\varphi(t)$ is of special interest for us. It is the unique solution of the linear differential equation

$$\varphi'(t) = \rho\varphi(t) + a(t), \quad \varphi(0) = 0.$$

The intuitive interpretation of this equation is clear. The total increment of $\varphi(t)$ during the time interval $[t, t+dt)$, which is $\varphi'(t)dt$, is formed by the sum of two different components: one, which is $\rho\varphi(t)dt$, is the change due to the interest rate ρ; and the other, $a(t)dt$, is the new payment. Suppose now that several payments, say k of them, are made with regard to several financial obligations, or paid into several different accounts, at possibly distinct rates $\rho_1, \ldots, \rho_k$. In this case we will have k equations

$$\varphi_1'(t) = \rho_1\varphi_1(t) + a_1(t),$$
$$\cdots \tag{14.7}$$
$$\varphi_k'(t) = \rho_k\varphi_k(t) + a_k(t),$$

which can be more conveniently written as a single vector equation

$$\varphi'(t) = \mathbf{D}\varphi(t) + \mathbf{a}(t), \tag{14.8}$$

where $\varphi(t)$ and $\mathbf{a}(t)$ are vector-functions

$$\varphi(t) = \begin{pmatrix} \varphi_1(t) \\ \vdots \\ \varphi_k(t) \end{pmatrix}, \qquad \mathbf{a}(t) = \begin{pmatrix} a_1(t) \\ \vdots \\ a_k(t) \end{pmatrix},$$

and $\mathbf{D}$ is a $k \times k$ diagonal matrix

$$\mathbf{D} = \begin{pmatrix} \rho_1 & 0 & \cdots & 0 \\ 0 & \rho_2 & \cdots & 0 \\ \cdots & \cdots & \cdots & \cdots \\ 0 & 0 & \cdots & \rho_k \end{pmatrix}.$$

Let us now consider the same equation, but with a more general matrix

$$\mathbf{R} = \begin{pmatrix} \rho_{11} & \rho_{12} & \cdots & \rho_{1k} \\ \rho_{21} & \rho_{22} & \cdots & \rho_{2k} \\ \cdots & \cdots & \cdots & \cdots \\ \rho_{k1} & \rho_{k2} & \cdots & \rho_{kk} \end{pmatrix},$$

on the right side of (14.8), so that

$$\varphi'(t) = \mathbf{R}\varphi(t) + \mathbf{a}(t), \tag{14.9}$$

or

$$\varphi_1'(t) = \rho_{11}\varphi_1(t) + \rho_{12}\varphi_2(t) + \cdots + \rho_{1k}\varphi_k(t) + a_1(t),$$
$$\cdots \tag{14.10}$$
$$\varphi_k'(t) = \rho_{k1}\varphi_1(t) + \rho_{k2}\varphi_2(t) + \cdots + \rho_{kk}\varphi_k(t) + a_k(t).$$

In addition to equation (14.7), in equations (14.10) it is now assumed that the ρ_{ij} portion from the j-th account is transferred into the i-th account; that is, the redistribution within accounts is now possible.

It seems very natural to write the solution of equation (14.9), with initial condition $\varphi(0) = 0$, in the form

$$\varphi(t) = \int_0^t e^{(t-s)\mathbf{R}} \mathbf{a}(s) \, ds. \tag{14.11}$$

And indeed this is the correct solution, provided we agree on the meaning of the expression

$$e^{(t-s)\mathbf{R}}.$$

Is this exponential a matrix, and if so, what kind of matrix? In fact, it is a matrix, defined through the Taylor expansion for e^x,

$$e^x = 1 + \sum_{i=1}^{\infty} \frac{x^i}{i!}.$$

Namely,

$$e^{(t-s)\mathbf{R}} = I + \sum_{i=1}^{\infty} \frac{(t-s)^i}{i!} \mathbf{R}^i.$$

Here I is a unit matrix

$$I = \begin{pmatrix} 1 & 0 & \cdots & 0 \\ 0 & 1 & \cdots & 0 \\ \cdots & \cdots & \cdots & \cdots \\ 0 & 0 & \cdots & 1 \end{pmatrix},$$

and $\mathbf{R}^i$ is the i-th power of $\mathbf{R}$, namely, $\mathbf{R}^i = \mathbf{R} \cdot \mathbf{R}^{i-1}$. Therefore, for an arbitrary vector $\mathbf{z}$ the vector $e^{(t-s)\mathbf{R}} \mathbf{z}$ is defined as

$$e^{(t-s)\mathbf{R}} \mathbf{z} = \mathbf{z} + \sum_{i=1}^{\infty} \frac{(t-s)^i}{i!} \mathbf{R}^i \mathbf{z}.$$

In the case of a diagonal matrix $\mathbf{D}$ the expression for $\mathbf{D}^i$ becomes very simple:

$$\mathbf{D}^i = \begin{pmatrix} \rho_1^i & 0 & \cdots & 0 \\ 0 & \rho_2^i & \cdots & 0 \\ \cdots & \cdots & \cdots & \cdots \\ 0 & 0 & \cdots & \rho_k^i \end{pmatrix},$$

so that

$$\mathbf{D}^i \mathbf{z} = \begin{pmatrix} \rho_1^i z_1 \\ \vdots \\ \rho_k^i z_k \end{pmatrix},$$

and the vector $e^{(t-s)\mathbf{D}} \mathbf{z}$ has coordinates

$$\left(1 + \sum_{i=1}^{\infty} \frac{(t-s)^i}{i!} \rho_1^i \right) z_1 = e^{(t-s)\rho_1} z_1,$$

$$\cdots$$

$$\left(1 + \sum_{i=1}^{\infty} \frac{(t-s)^i}{i!} \rho_k^i\right) z_k = e^{(t-s)\rho_k} z_k.$$

The application of the matrix $e^{(t-s)\mathbf{D}}$ to the vector $\mathbf{a}(s)$ amounts to the multiplication of its coordinates by $e^{(t-s)\rho_1}, \ldots, e^{(t-s)\rho_k}$ respectively, and the solution of (14.8) splits into k known solutions:

$$\int_0^t e^{(t-s)\mathbf{D}} \mathbf{a}(s)\,ds = \begin{pmatrix} \int_0^t e^{(t-s)\rho_1} a_1(s)\,ds \\ \cdots \\ \int_0^t e^{(t-s)\rho_k} a_k(s)\,ds \end{pmatrix}. \tag{14.12}$$

Now let $\mathbf{R}$ be a general matrix. As we know, see, e.g., Glazman and Ljubich [1974], the vector $\mathbf{z}$ is called an eigenvectorof the matrix $\mathbf{R}$, corresponding to the eigenvalue ρ, if

$$\mathbf{R z} = \rho \mathbf{z}.$$

Any matrix has at least one eigenvector and a corresponding eigenvalue. We will, however, restrict ourselves to matrices that have a complete eigenbasis, that is, matrices having k linearly independent eigenvectors. Matrices of this type, and the corresponding linear operators in k-dimensional space, are called scalar type matrices (or operators). This is a very general type of matrix, and in particular includes all nonsingular or invertible matrices, although not all matrices are of scalar type.

If $\mathbf{z}$ is an eigenvector of a matrix $\mathbf{R}$, with corresponding eigenvalue ρ, then

$$\mathbf{R}^i \mathbf{z} = \rho^i \mathbf{z},$$

and thus the matrix $e^{(t-s)\mathbf{R}}$ applied to the vector $\mathbf{z}$ produces the very simple result:

$$e^{(t-s)\mathbf{R}} \mathbf{z} = e^{(t-s)\rho} \mathbf{z}.$$

Since the matrix $\mathbf{R}$ has an eigenbasis $\mathbf{z}_1, \ldots, \mathbf{z}_k$, one can expand the vector $\mathbf{a}(s)$, for every s, in this basis:

$$\mathbf{a}(s) = \sum_{j=1}^{k} \zeta_j(s) \mathbf{z}_j. \tag{14.13}$$

Consequently

$$e^{(t-s)\mathbf{R}}\mathbf{a}(s) = \sum_{j=1}^{k} \zeta_j(s) e^{(t-s)\rho_j} \mathbf{z}_j,$$

and the function $\varphi(t)$ in (14.11) assumes a very simple form:

$$\varphi(t) = \sum_{j=1}^{k} \mathbf{z}_j \int_0^t e^{(t-s)\rho_j} \zeta_j(s) \, ds. \qquad (14.14)$$

In financial applications, one can interpret the eigenvectors $\mathbf{z}_j$ as new "accounts". In between these accounts there is no more redistribution and each of them grows at rate ρ_j, $j = 1, \ldots, k$. Note that these rates, viz. the eigenvalues of the matrix $\mathbf{R}$ corresponding to different eigenvectors, are not necessarily distinct.

◇ **Exercise.** What are the eigenvectors and eigenvalues, and what is the expansion (14.13), in the case of the diagonal matrix $\mathbf{D}$? Do we again obtain (14.12) from (14.14)? △

◇ **Exercise.** Consider a specific matrix $\mathbf{R}$:

$$\mathbf{R} = \begin{pmatrix} 2 & 1 \\ 1 & 2 \end{pmatrix}.$$

This means that we have only two accounts of sizes, or of values, $\varphi_1(t)$ and $\varphi_2(t)$ which satisfy the equations

$$\varphi_1'(t) = 2\varphi_1(t) + \varphi_2(t) + a_1(t),$$
$$\varphi_2'(t) = \varphi_1(t) + 2\varphi_2(t) + a_2(t).$$

First assume that the initial condition $\varphi_1(0) = c_1, \varphi_2(0) = c_2$ is non-zero, while there are no payments: $a_1(t) = a_2(t) = 0$ for all t.
a) Is it easy to guess from the equations what the behavior of $\varphi(t)$ is in this situation? For example, do they have the same asymptotic as $t \to \infty$?
b) Show that the solution has the form

$$\varphi(t) = e^{t\mathbf{R}} \begin{pmatrix} c_1 \\ c_2 \end{pmatrix}$$

and that explicit form of this solution is

$$\varphi_1(t) = \frac{c_1 + c_2}{2} e^{3t} + \frac{c_1 - c_2}{2} e^t,$$

$$\varphi_2(t) = \frac{c_1 + c_2}{2} e^{3t} - \frac{c_1 - c_2}{2} e^t.$$

c) Assume now that $c_1 = c_2 = 0$ and that there are constant pay-ments $a_1(t) = a_2(t) = 1$ for all t. What is the form of the solution in this case? △

Right-tail behavior of $\widehat{F}_n$. Non-parametric confidence bounds for expected remaining life

Consider the expression for the actuarial value of a continuously payable premium at constant rate p with discounting, which we derived in (12.4):

$$p \int_x^\infty e^{-\rho(y-x)} \frac{1-F(y)}{1-F(x)} dy.$$

For small values of x and advanced ages y, say for $x = 25$ and y exceeding 80 years, the values of $1 - F(y)$ seem to play a minor role in this expression, especially when one considers the discounting taking place within the integral; and the problem of accurate estimation of F at high ages does not seem very important.

In practice, however, it rarely happens that the premium established at younger ages remains unchanged until the end of life. Insurance companies, especially life insurance and health insurance companies, always revise, or feel obliged to revise, premiums over time.

Supposing premiums to be reviewed every 3 to 5 years, the accurate estimation of F for the advanced ages becomes important not only for insurers and insureds, but also for those who shape social security policies for the elderly.

If n is large, then the empirical process V_n, discussed in Lectures 3–5, is still well approximated by the Brownian bridge at quite high ages x. This can happen when samples include sufficiently many elderly people, say if there are several hundred people in the $85 - 95$ age group. Much more than this number can be found on national scales. For example, in the New Zealand population there are about 31000 men of age 90 or over.

According to its definition in Lecture 4,

$$V_n(x) = \frac{1}{\sqrt{n}} [z_n(x) - nF(x)]$$

where, as we know, $z_n(x)$ is a binomial random variable with parameters n and $F(x)$; it is more natural in the present context to rewrite $V_n(x)$ as

$$V_n(x) = -\frac{1}{\sqrt{n}} [n - z_n(x) - n(1 - F(x))]$$

and to speak about the number of survivals $n - z_n(x)$, which is a binomial random variable with parameters n and $1 - F(x)$. If now $n(1 - F(x))$ is of the order of 100 or more, or even as low as 50, then the distribution of $n - z_n(x)$ is still well approximated by the normal distribution; see the de Moivre–Laplace theorem (4.2).

In specialized portfolios and in many clinical trials, though, it is rare to have large n. In these cases $n(1 - F(x))$ for large x often turns out to be of the order of, say, less than twenty or so. In such situations the normal approximation for $V_n(x)$ cannot be used, and certainly the normal approximation for the empirical process "as a whole", on the entire right tail, will not be valid. Thus, we need to develop a different asymptotic theory applicable for such cases.

And so, let us consider the right "tail" of the distribution function F on the set of values $\{x \geq x_{0n}\}$, where the boundary x_{0n} increases with n so that

$$n(1 - F(x_{0n})) \longrightarrow c,$$

in which c is a positive constant. Consider the tail process $n - z_n(x), x \geq x_{0n}$. There is an intimate connection between this process and Poisson processes, and there are several ways to explore this connection.

Recall that increments of a Poisson process on non-overlapping intervals are independent Poisson random variables; and further recall that for a time homogeneous Poisson process, with constant intensity

μ, the distribution of each increment is

$$P\{\xi(t+\Delta t) - \xi(t) = k\} = \frac{(\mu\Delta t)^k}{k!} e^{-\mu\Delta t}, \quad k = 0, 1, 2, \ldots.$$

For the standard Poisson process the parameter $\mu = 1$.

Now consider the tail process in the new time $t = n(1 - F(x))$; that is, consider

$$\xi_n(t) = n\left(1 - \widehat{F}_n(x)\right) = n - z_n(x), \quad t = n(1 - F(x)).$$

Unlike the range of ages $x \in [x_{0n}, \infty)$, which now changes with n, the range of t stays the same: $t \in [0, c]$. There is also a certain inversion: small values of t correspond to large values of x. We could avoid this inversion by considering $n(\widehat{F}_n(x) - \widehat{F}_n(x_{0n})) = z_n(x) - z_n(x_{0n})$ and studying it in time $s = n(F(x) - F(x_{0n}))$, or similar. However, for the time being we proceed as we started.

Let us formulate a Poisson limit theorem for ξ_n:

if $n \to \infty$ and $x_{0n} \to \infty$ so that $n(1 - F(x_{0n})) \to c$, then

$$\xi_n \xrightarrow{d_f} \xi, \tag{15.1}$$

where $\xi(t)$, $0 < t \le c$, is a standard Poisson process.

Considered on the total range $0 < t < n$, the process ξ_n is a binomial process based on n independent random variables, uniformly distributed on $[0, n]$. Therefore here we study the "small" portion of it, restricted to a "neighborhood" $[0, c]$ of 0: for, say, $c = 10$ and $n = 1000$, the interval $[0, 10]$ can be considered a neighborhood of 0 relative to the interval $[0, 1000]$.

The proof of convergence of the finite-dimensional distributions (15.1) is offered below as an exercise. In contrast to the case of limiting Brownian motion, for the Poisson process it is possible to show (see, e.g., Reiss and Thomas [2007]), that for statistics $\psi(\xi_n)$ of the type

$$\sup_{0 < t \le c} |\xi_n(t) - t| = \sup_{x \ge x_{0n}} |z_n(x) - nF(x)|,$$

i.e. the statistics, which depend on values of $\xi_n(t)$ at infinitely many t, just the finite-dimensional convergence implies that the limit theorem

$$\psi(\xi_n(t),\ t \le c) \xrightarrow{d} \psi(\xi(t),\ t \le c)$$

is also valid. Let us see how we can use this limit theorem.

At the tail $\{x \ge x_{0n}\}$, both values of $1 - F(x)$ and $1 - \widehat{F}_n(x)$ are small. Hence, it is more meaningful to speak about the relative deviation of $1 - \widehat{F}_n(x)$ from $1 - F(x)$, instead of the absolute deviation $\widehat{F}_n(x) - F(x)$. Consider

$$\frac{1 - \widehat{F}_n(x)}{1 - F(x)} = \frac{n - z_n(x)}{n(1 - F(x))} = \frac{\xi_n(t)}{t}.$$

Since ξ_n remains random, no matter how big n is, our immediate observation is that on the tail $[x_{0n}, \infty)$ the ratio $[1 - \widehat{F}_n(x)]/[1 - F(x)]$ does not converge to 1, and in this sense $1 - \widehat{F}_n(x)$ is not a consistent estimator for $1 - F(x)$. From (15.1) it follows that

$$P\left\{ \sup_{x \ge x_{0n}} \frac{1 - \widehat{F}_n(x)}{1 - F(x)} < \lambda \right\} \longrightarrow P\left\{ \sup_{0 < t \le c} \frac{\xi(t)}{t} < \lambda \right\}. \tag{15.2}$$

This latter probability we can calculate easily. Thus, we still have the possibility to speak about how far $1 - \widehat{F}_n(x)$ can deviate relative to $1 - F(x)$.

More specifically, the probability in the right-hand side of (15.2) is the probability that the Poisson process $\xi(t)$ will not cross the boundary λt on the whole interval $(0, c]$. The calculation of non-crossing probabilities for various boundaries has been studied by many authors; see, e.g., references in Shorack and Wellner [2009], ch. 8.3, and in Khmaladze and Shinjikashvili [2001]. In the latter paper a method was given to calculate these probabilities for more or less arbitrary boundaries and for values of c up to several thousands. As a matter of fact, the approach allows one to calculate non-crossing probabilities for finite n as well, but the rate of convergence in (15.1) is good and calculations with the same c but finite n often seem unnecessary.

As soon as we know the distribution of the statistic

$$\sup_{0<t\leq c} [\xi_n(t)/t],$$

exactly or with good accuracy, we can obtain confidence intervals for the tail $1 - F(x)$. Namely, if we choose λ so that

$$P\left\{ \sup_{0<t\leq c} \frac{\xi(t)}{t} < \lambda \right\} = 1 - \alpha,$$

i.e., if

$$P\left\{ 1 - \widehat{F}_n(x) \leq \lambda(1 - F(x)), \text{ for all } x \geq x_{0n} \right\} \approx 1 - \alpha,$$

then

$$\frac{1}{\lambda}\left[1 - \widehat{F}_n(x)\right]$$

will be the lower confidence bound for $1 - F(x)$, and the choice of λ does not depend on the distribution function F: the bound is non-parametric.

It is worth noting that there is some freedom of choice in finding confidence bounds for $1 - F(x)$. Although for the simple bound like $\lambda(1 - F(x))$ in (15.2), which translates into a linear bound λt for the Poisson process, many things are known analytically (see below), this bound is not the only one possible and, perhaps, not the best. Indeed, instead of the supremum employed in (15.2), we can, for example, consider the distribution of the different supremum:

$$P\left\{ \sup_{0<t\leq c} \frac{\xi(t)}{t+\varepsilon} < \lambda \right\}, \tag{15.3}$$

which leads to the bound:

$$\frac{1}{\lambda}\left[1 - \widehat{F}_n(x) - \varepsilon/n\right] < 1 - F(x), \text{ for all } x > x_{0n}. \tag{15.4}$$

Already for $\varepsilon = 0.75$ or $\varepsilon = 1$ the distribution function (15.3) is far more "concentrated" than the distribution function in the right-hand side of (15.2). In particular, the value of λ corresponding to the level $1 - \alpha = 0.95$ is essentially smaller than it was for the distribution in (15.2). Because of this, although the square bracket in (15.4) is somewhat smaller than before, the bound itself may increase for the majority of the values of x, and therefore represent an improvement.

As an illustration we present here the table of values of λ corresponding to confidence levels $1 - \alpha = 0.9$ and $1 - \alpha = 0.95$. The value of λ for non-zero ε is essentially smaller than for $\varepsilon = 0$.

$c = 20$	$\varepsilon = 0.95$	$\lambda = 4$	probability $= 0.95$
$c = 20$	$\varepsilon = 0.75$	$\lambda = 4$	probability $= 0.90$
$c = 10$	$\varepsilon = 0.95$	$\lambda = 4$	probability $= 0.95$
$c = 10$	$\varepsilon = 0.75$	$\lambda = 4$	probability $= 0.90$
$c = 20$	$\varepsilon = 0$	$\lambda = 20$	probability $= 0.95$
$c = 20$	$\varepsilon = 0$	$\lambda = 10$	probability $= 0.90$

Table 15.1 *Values of λ calculated by E. Shinjikashvili for this lecture*

◇ **Exercise.** Let $h(t)$ be a continuous increasing function. As was mentioned above, it is easy to calculate the non-crossing probability

$$P\{\xi(t) < \lambda h(t), \text{ for all } 0 < t \leq c\}.$$

But then

$$\max\left(0, \frac{1}{n} h^{-1}\left(\frac{1}{\lambda}[n - z_n(x)]\right)\right), \quad x \geq x_{0n},$$

is the bound for $1 - F(x)$, $x \geq x_{0n}$. Verify this claim.   △

◇ **Exercise.** To prove the convergence (15.1) for any integer k and for any collection of points $0 = t_0 < t_1 < \cdots < t_k = c < t_{k+1} = n$, consider the vector of increments $\Delta\xi_n(t_j) = \xi_n(t_{j+1}) - \xi_n(t_j)$, $j = 0, \ldots, k$, and, using Stirling's formula, derive the limit of their joint distribution, given in (4.5). This should lead to the limit theorem

$$\left(\Delta\xi_n(t_j), \ j = 0, \ldots, k-1\right) \xrightarrow{d} \left(\Delta\xi(t_j), \ j = 0, \ldots, k-1\right),$$

which is equivalent to the limit theorem

$$\left(\xi_n(t_j),\ j=0,\ldots,k\right) \xrightarrow{d} \left(\xi(t_j),\ j=0,\ldots,k\right).$$

Note the role of the last increment $\xi_n(n) - \xi_n(c) = z_n(x_{0n})$, although it does not appear in the previous display.   △

Let us now consider another connection between the tail of $1 - \widehat{F}_n$ and Poisson processes. It will not be a limit theorem, but an exact equality. For variety's sake consider now the process $z_n(x) - z_n(x_{0n})$, $x \geq x_{0n}$, instead of $n - z_n(x)$, $x \geq x_{0n}$. We split what we want to say into two statements.

First, let $\tilde{z}_L(x), x \geq x_{0n}$, denote a binomial process, based on L independent, identically distributed random variables, each with the distribution function

$$\tilde{F}(x) = F(x|x_{0n}) = \frac{F(x) - F(x_{0n})}{1 - F(x_{0n})},$$

so that $\tilde{z}_L(x_{0n}) = 0$ and $\tilde{z}_L(\infty) = L$. Then we have:

the conditional distribution of the process $z_n(x) - z_n(x_{0n})$, $x \geq x_{0n}$, under the condition that $n - z_n(x_{0n}) = L$, is the same as the distribution of the process $\tilde{z}_L(x), x \geq x_{0n}$;

in particular, for any boundaries $\tilde{g}(x)$ and $\tilde{h}(x)$,

$$\mathsf{P}\left\{\tilde{g}(x) < z_n(x) - z_n(x_{0n}) < \tilde{h}(x),\ \text{for all } x \geq x_{0n} \,\middle|\, n - z_n(x_{0n}) = L\right\}$$

$$= \mathsf{P}\left\{\tilde{g}(x) < \tilde{z}_L(x) < \tilde{h}(x),\ \text{for all } x \geq x_{0n}\right\}.$$

Although its proof is very simple, this is a remarkably useful statement. It says that as soon as we know that $z_n(x_{0n}) = n - L$, so that there are L observations in the tail, the future evolution of the binomial process depends on nothing else from the past and follows the conditional distribution $\tilde{F}$.

To prove this statement, again use increments rather than the values of the processes themselves. For any finite k and any collection of points $x_{0n} < x_1 < \cdots < x_{k-1} < x_k = \infty$ consider the conditional joint distribution of the differences $\Delta(z_n(x_j) - z_n(x_{0n})) = \Delta z_n(x_j)$: for $\sum l_j = L$

$$\frac{P\{\Delta z_n(x_j) = l_j, j = 0, \ldots, k-1, z_n(x_{0n}) = n - L\}}{P\{z_n(x_{0n}) = n - L\}}$$

$$= \frac{n!}{(n-L)!\Pi_{j=0}^{k-1}l_j!}\Pi_{j=0}^{k-1}[\Delta F(x_j)]^{l_j}[F(x_{0n})]^{n-L}$$

$$\times \frac{(n-L)!L!}{n!}\frac{1}{[F(x_{0n})]^{n-L}[1-F(x_{0n})]^L}$$

$$= \frac{L!}{\Pi_{j=0}^{k-1}l_j!}\Pi_{j=0}^{k-1}\Big[\frac{\Delta F(x_j)}{1-F(x_{0n})}\Big]^{l_j},$$

and the last expression is just the probability

$$P\{\Delta \tilde{z}_L(x_j) = l_j, j = 0, \ldots, k-1\}.$$

This proves the first claim, and, therefore, the second as well.

Let now $\eta(x), x \geq x_{0n}$, denote a Poisson process with expected value

$$E\eta(x) = \tilde{\lambda}\tilde{F}(x),$$

although $\tilde{\lambda}$ will play below no role: it merely helps in an intuitive understanding of what is needed from η; that is, we do not need to insist that $E\eta(\infty) = 1$, it can be any $\tilde{\lambda}$.

Recall again that increments of η on disjoint intervals are independent Poisson random variables, and the distribution of each increment is

$$P\{\eta(x+\Delta x) - \eta(x) = k\} = \frac{(\tilde{\lambda}\Delta\tilde{F}(x))^k}{k!}e^{-\tilde{\lambda}\Delta\tilde{F}(x)}, \quad k = 0, 1, 2, \ldots,$$

where $\Delta\tilde{F}(x) = \tilde{F}(x+\Delta x) - \tilde{F}(x)$. Then we have:

the distribution of the process $\tilde{z}_L(x), x \geq x_{0n}$, is the same as the conditional distribution of the Poisson process $\eta(x), x \geq x_{0n}$, under the condition that $\eta(\infty) = L$;

in particular, for any boundaries $\tilde{g}(x)$ and $\tilde{h}(x)$,

$$P\{\tilde{g}(x) < \tilde{z}_L(x) < \tilde{h}(x), \text{ for all } x \geq x_{0n}\}$$

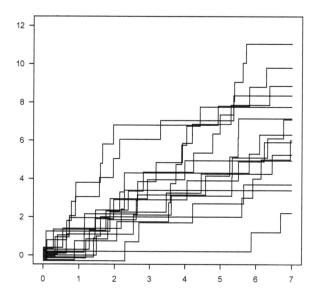

Figure 15.1 *Several trajectories of a standard Poisson process on the interval*
$[0,7]$, unrestricted by their last value.

$$= P\left\{\tilde{g}(x) < \eta(x) \le \tilde{h}(x), \text{ for all } x \ge x_{0n} \mid \eta(\infty) = L\right\}.$$

This is also a very interesting and telling statement. It says that as
soon as we know, and fix, the total number L of jumps of a Poisson
process, looking backwards, its evolution is exactly the same as that
of a binomial process based on L independent random variables with
distribution function

$$\frac{E\eta(x)}{E\eta(\infty)} = \tilde{F}(x)$$

(and $\tilde{\lambda}$ cancels out).

The new insight we have gained now is not the form of the process
$z_n(x) - z_n(x_{0n})$, but the idea that we can also use conditional Poisson
non-crossing probabilities. Returning to the process ξ_n we can see that
the distribution of the process $\xi_n(t), 0 < t < c$, under the condition

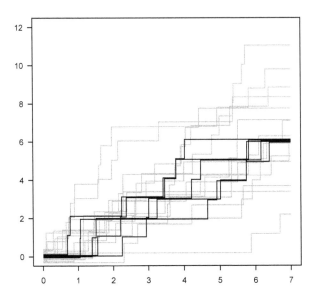

Figure 15.2 *Several trajectories of a standard Poisson process, which all reach the same number L = 6 at a given point c = 7. Imagine a bundle of all such trajectories with one fixed L. If we consider only trajectories from this bundle, their distribution will not be Poisson, but that of a binomial process. In the background, in gray, the unrestricted trajectories are redrawn from Figure 15.1.*

$\xi_n(c) = L$, *is the same as the distribution of a standard Poisson process* $\xi(t), 0 < t < c$, *under the condition* $\xi(c) = L$; *both are equal to the distribution of a uniform binomial process on* $[0, c]$ *with L observations.*

Therefore we can either calculate the probability

$$P\left\{1 - \widehat{F}_n(x) < \lambda(1 - F(x)) \text{ for all } x > x_{0n}\right\},$$

or, more generally, the probability

$$P\left\{1 - \widehat{F}_n(x) < \lambda h(1 - F(x)) \text{ for all } x > x_{0n}\right\}$$

$$= \mathsf{P}\{\xi_n(t) < \lambda n h(t) \text{ for all } t \in [0,c]\}$$

using its approximation

$$\mathsf{P}\{\xi(t) < \lambda n h(t) \text{ for all } t \in [0,c]\},$$

and find λ such that the latter probability equals the desired confidence probability $1 - \alpha$; or we can calculate the conditional probability, using the equality

$$\mathsf{P}\left\{1 - \widehat{F}_n(x) < \lambda h(1 - F(x)) \text{ for all } x > x_{0n} \mid 1 - \widehat{F}_n(x_{0n}) = L/n\right\}$$

$$= \mathsf{P}\{\xi(t) < \lambda n h(t) \text{ for all } t \in [0,c] \mid \xi(c) = L\} \qquad (15.5)$$

and again find λ_L, such that this probability equals the confidence level $1 - \alpha$. Often $\lambda_L < \lambda$ and therefore the confidence bound

$$\frac{1}{\lambda_L} h^{-1}(1 - \widehat{F}_n(x))$$

is higher and, hence, better. However, one has to have observations in order to know L, while calculation of λ does not require anything. Note again, that neither λ_L nor λ depends on F — they remain the same for all continuous F.

◇ **Exercise.** Again, using increments of η between points x_j of any given partition, prove the previous proposition. In spelling out the conditional joint distribution of the increments do not forget the last increment $\eta(\infty) - \eta(x_{k-1})$ in the numerator. △

◇ **Exercise.** a) If $\xi(t), 0 < t < c$, is a standard Poisson process, then its conditional distribution given $\xi(c) = L$ is the same as the distribution of the uniform binomial process based on L independent random variables, uniformly distributed on the interval $[0,c]$. Verify this statement.
b) Verify the same statement for the conditional distribution of the process ξ_n, considered on the previous pages, under the condition $\xi_n(c) = L$. You may be sufficiently convinced by considering only the one-dimensional conditional probability

$$\frac{P\{\xi_n(t) = k, \xi_n(c) - \xi_n(t) = L - k\}}{P\{\xi(c) = L\}}$$

and the same probability for ξ. They both should be equal to

$$P\{z_L(t) = k\} = \frac{L!}{k!(L-k)!} \left(\frac{t}{c}\right)^k \left(1 - \frac{t}{c}\right)^{L-k}.$$

$\triangle$

◇ **Exercise.** Show that if $\xi(t), 0 < t < c$, is a standard Poisson process, then $\tilde{\xi}(s) = \xi(c) - \xi(c - s)$ is also a standard Poisson process. $\triangle$

What we have learned so far leads also to the following interesting result – to non-parametric confidence bounds for the expected remaining life,

$$E[T - x_{0n} \mid T > x_{0n}] = \frac{\int_{x_{0n}}^{\infty} [1 - F(x)] \, dx}{1 - F(x_{0n})}.$$

If now

$$\lambda \frac{1 - F(x)}{1 - F(x_{0n})} > \frac{1 - \widehat{F}_n(x)}{1 - \widehat{F}_n(x_{0n})} \quad \text{for all } x > x_{0n}, \tag{15.6}$$

then, obviously,

$$E[T - x_{0n} \mid T > x_{0n}] > \frac{1}{\lambda} \frac{\int_{x_{0n}}^{\infty} [1 - \widehat{F}_n(x)] \, dx}{1 - \widehat{F}_n(x_{0n})}. \tag{15.7}$$

However, inequality (15.6) is equivalent to the inequality

$$\lambda [n - z_n(x_{0n})] \frac{1 - F(x)}{1 - F(x_{0n})} > n - z_n(x)$$

or

$$\xi_n(t) < \lambda \frac{t}{c} \xi_n(c), \quad \text{for all } t \leq c.$$

The conditional probability of this inequality is known, see Ishii [1959] and Smirnov [1961]:

$$P\left\{ \xi_n(t) < \lambda \frac{t}{c} \xi_n(c) \text{ for all } t \in [0, c] \mid \xi_n(c) = L \right\} = \frac{\lambda}{1 + \lambda},$$

and is independent of L. Then finally

$$\mathrm{P}\left\{E[T - x_{0n} \mid T > x_{0n}] > \frac{1}{\lambda} \frac{\int_{x_{0n}}^{\infty} [1 - \widehat{F}_n(x)]\,dx}{1 - \widehat{F}_n(x_{0n})}\right\} \geq \frac{\lambda}{1+\lambda}.$$

We do not have equality here, because (15.6) is only a sufficient condition for the inequality (15.7).

◇ **Exercise.** Use the recurrence given in the section 15.2s to conduct a computer experiment trying various functions h in (15.5). The "higher" the resulting confidence limit for $1 - F(x)$, the better the bound for $E[T - x_{0n} \mid T > x_{0n}]$.   △

15.1* Asymptotic form of the excess life distribution

Consider again the conditional distribution $\tilde{F}$, which was evoked in the analysis of the tail process $z_n(x) - z_n(x_{0n})$, $x > x_{0n}$. More precisely, consider it in a centered and scaled form, which we again denote by $\tilde{F}$ adding only a subscript n:

$$\tilde{F}(x_{0n} + a_n y) = \tilde{F}_n(y) = \frac{F(x_{0n} + a_n y) - F(x_{0n})}{1 - F(x_{0n})}.$$

This $\tilde{F}_n$ is the distribution function of the scaled remaining life $(T - x_{0n})/a_n$, obviously under the condition $T > x_{0n}$. If, when T exceeds a high level x_{0n}, it tends be closer to x_{0n}, the difference $T - x_{0n}$ will need to be scaled up and a_n will need to decrease to 0 with n. If, as soon as it exceeds x_{0n}, T tends to exceed it by large amount, then $T - x_{0n}$ will need to be scaled down and $a_n \to \infty$. It may be also that there is no need for any scaling, in which case a_n can be chosen as a constant. We will have examples of all three situations below.

According to the Balkema–de Haan–Pickands theorem (see, e.g., de Haan and Ferreira [2006]), if there exists a sequence of normalizing constant a_n, such that $\tilde{F}_n(y)$ converges to some non-degenerate distribution $G(y)$,

$$\tilde{F}_n(y) \to G(y), \quad y > 0,$$

then this G can only be of the following form:

$$G(y) = \begin{cases} 1 - (1 + \theta y/\sigma)^{-1/\theta}, \text{ for } \theta > 0, \\ 1 - e^{-y/\sigma}, \text{ for } \theta = 0. \end{cases}$$

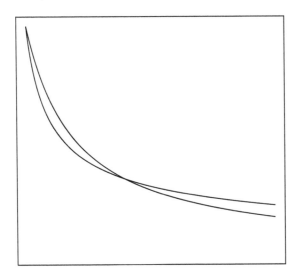

Figure 15.3 *Two graphs of the tail of the Pareto distribution function with different parameters. Which one is better?*

The use of this parametric form can give us the following advantage. So far, the non-parametric confidence bounds that we obtained for $1 - F(x)$ have been such that they were equal to 0 at $x > \max T_i$ or earlier, thus not allowing researchers to penetrate beyond the largest observation. However, approximation of $1 - F(x)$ by $c(1 - G(y))/n$ produces a natural parametrization of $1 - F(x)$. Although the confidence bound with two parameters again may look somewhat awkward, cf. Figure 15.3, reasonable estimation of parameters will produce a reasonable approximation of the tail $1 - F(x)$ beyond the largest observation $\max T_i$, i.e. in the range where there are no observations.

The assumption that $\tilde{F}_n$ converges as $n \to \infty$ is a mild assumption, but is still an assumption. Let us see how is it satisfied for the distributions useful for analysis of lifetimes.

First note, cf. (1.14), that

$$1 - F(x_{0n}) = \exp\left[-\int_0^{x_{0n}} \mu(z)dz\right]$$

and

$$1 - F(x_{0n} + a_n y) = \exp\left[-\int_0^{x_{0n} + a_n y} \mu(z)dz\right]$$

and, therefore, for the tail $1 - \tilde{F}_n(y)$ one obtains

$$1 - \tilde{F}_n(y) = \exp\left[-\int_{x_{0n}}^{x_{0n} + a_n y} \mu(z)dz\right]. \tag{15.8}$$

In all expressions here μ is the force of mortality. If the asymptotic relation

$$\int_{x_{0n}}^{x_{0n} + a_n y} \mu(z)dz \sim \mu(x_{0n})a_n y, \quad x_{0n} \to \infty, \tag{15.9}$$

holds, the only choice for a_n is $a_n \sim c/\mu(x_{0n})$, with c a generic notation for constant, and the only possible limit for $\tilde{F}_n$ is the exponential distribution:

$$1 - \tilde{F}_n(y) \to \exp\left[-cy\right]. \tag{15.10}$$

In particular, suppose F is a gamma distribution function, defined in (2.9). In Lecture 2 we have shown that for this distribution $\mu(z) \to \lambda$ as $z \to \infty$. With $a_n = 1$ this immediately leads to (15.10) with $c = \lambda$.

For the logistic distribution again $\mu(x) \to k$, see (2.6), and therefore, again, one can take $a_n = 1$ and obtain $\exp[-ky]$ as the limit.

For the Weibull distribution function, using the explicit expression (2.3) we obtain:

$$1 - \tilde{F}_n(y) = \exp\left[(\lambda x_{0n})^k - (\lambda(x_{0n} + a_n y))^k\right]$$

and assuming a_n at least bounded,

$$(\lambda x_{0n})^k - (\lambda(x_{0n} + a_n y))^k = (\lambda x_{0n})^k\left[1 - (1 + \frac{a_n}{x_{0n}}y)^k\right]$$

$$\sim (\lambda x_{0n})^k\left(-k\frac{a_n}{x_{0n}}y\right) = -k\lambda^k x_{0n}^{k-1} a_n y.$$

This implies that a_n actually converges to 0 as c/x_{0n}^{k-1} and that G is again the exponential distribution function.

For the Gompertz distribution, with the force of mortality (2.4), the same is true again with $a_n \sim e^{-cx_{0n}}$.

Hence, for all distributions that we have considered, with asymptotically constant or increasing failure rates, the only possible limit for the excess life distribution $\tilde{F}_n$ is the exponential distribution.

◇ **Exercise.** It is very interesting to see what happens if the rate of mortality increases even more rapidly than exponential function. For example, try the function

$$\mu(x) = \exp\left(\exp(cx)\right),$$

with, say, $c = 1/30$, and calculate the integral in (15.9) with $a_n = 1/\mu(x_{0n})$ numerically. Although R. Brownrigg carried out this little experiment for x_{0n} up to 100 and $y \in [0, 10]$, and confirmed that the asymptotic expression in (15.9) is again valid, still do it yourself.   △

◇ **Exercise.** The assumption that there is an ultimate finite age $\omega < \infty$ changes the situation with the limit distribution. Consider, for instance, what happens if one chooses F as a beta distribution. According to what we have seen in Lecture 2 choose

$$\mu(x) = \frac{b}{\omega - x}, \quad b > 0.$$

What will be then the limit of $\tilde{F}_n$? Do not forget that $x_{0n} + a_n y \leq \omega$ and therefore $y \leq (\omega - x_{0n})/a_n$.   △

As the last example choose a mixture F considered in Lecture 12:

$$1 - F(x) = \int_0^\infty \exp\left[-\lambda \int_0^x \mu(y)dy\right] dH(\lambda).$$

We stress that this F is not a model for distribution of an individual lifetime, but for a mixture from a population of differently distributed lifetimes.

From the statement (12.20) about its force of mortality η and from (15.8) one can derive for the corresponding $\tilde{F}_n$ that

$$1 - \tilde{F}_n(y) \sim \exp\left[-\ln\int_0^{x_{0n}+a_n y}\mu(z)dz + \ln\int_0^{x_{0n}}\mu(z)dz\right]$$

$$= \frac{\int_0^{x_{0n}}\mu(z)dz}{\int_0^{x_{0n}+a_n y}\mu(z)dz}.$$

If F is a mixture of exponential distributions, with $\mu(z)$ constant, or if it is a mixture of the Weibull distributions, with $\mu(z) = \lambda z^{k-1}$, one has to choose a_n increasing as $c x_{0n}$ and

$$1 - \tilde{F}_n(y) = \frac{x_{0n}^k}{(x_{0n}+a_n y)^k} \to \frac{1}{(1+cy)^k}.$$

This is, of course, in very sharp contrast with what we have seen in the previous cases.

15.2* Recurrence formulae for calculation of non-crossing probabilities

As before, let ξ be a standard Poisson process on the interval $[0,c]$ and consider the probability that this process will not cross two given boundaries g and h:

$$P\{g < \xi < h\} = P\{g(t) < \xi(t) < h(t), \text{for all } t \in [0,c]\}. \quad (15.11)$$

First we notice that, without loss of generality, the boundaries g and h can be assumed non-decreasing. Since for any $f(t), t \in [0,c]$, the functions $\underline{f}(t) = \min_{s \geq t} f(s)$ and $\overline{f}(t) = \max_{s \leq t} f(s)$ are non-decreasing, this is a direct implication of the following statement:

for any process ξ with non-decreasing trajectories and, therefore, for any point process,

$$P\{g < \xi_n < h\} = P\{\overline{g} < \xi_n < \underline{h}\}.$$

To see that this is true, just draw a piece-wise constant trajectory, such as that of a Poisson process, and two arbitrary functions g and h; it will be then clear that if the trajectory does not cross these g and h, it will not cross $\overline{g}$ and $\underline{h}$ either.

Probability (15.11) is the probability that an infinite number of inequalities is true simultaneously. However, for point processes, this probability is equal to the probability of only finitely many inequalities

$$g(t_j) < \xi_n(t_j) < h(t_j)$$

being true at strategically chosen points $\{t_j\}_1^N, t_j \in [0,c]$; that is why it is possible to calculate (15.11) exactly. Let us first construct the set $\{t_j\}_1^N$.

For a non-decreasing function $g(t), t \in [0,c], g(0) < 0$, and for any integer $j, 0 \le j \le g(c)$, let

$$x_j = \inf\{t : g(t) \ge j\}.$$

For a non-decreasing function $h(t), t \in [0,c], h(0) \ge 0$, and for any integer $j, h(0) \le j \le h(c)$, let

$$y_j = \sup\{t : h(t) \le j\}.$$

Denote by $\mathbf{X}(g)$ the set of all x_js and by $\mathbf{Y}(h)$ the set of all y_js. We need to add the point $t = c$ to these sets in what follows:

for any point process ξ and non-decreasing boundaries g and h

$P\{g < \xi < h\}$
$\quad = P\{g(t) < \xi(t) < h(t) \text{ for all } t \in \mathbf{X}(g) \cup \mathbf{Y}(h) \cup \{c\}\};$ (15.12)

for the upper boundary only

$$P\{\xi < h\} = P\{\xi(t) < h(t) \ \text{ for all } \ t \in \mathbf{Y}(h) \cup \{c\}\}.$$

Thus, once again, our statement shows that the event $\{g < \xi < h\}$ seemingly concerning infinitely many $\xi(t)$, for a point process, is only an event concerning finitely many of them. The number of $\xi(t_j)$'s involved is determined by the boundaries.

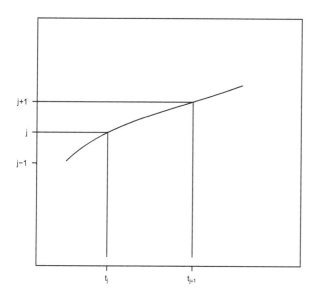

Figure 15.4 *If a trajectory of ξ starts at t_j on $j-1$ or below, and at t_{j+1} arrives to a value not higher that j, then in between it cannot cross h.*

If functions g and h have discontinuities, or are not strictly increasing, it may happen that several points in $\mathbf{X}(g)$ and in $\mathbf{Y}(h)$ are equal. However, if, say, h is continuous and increasing, then the y_js are all different, and

$$P\{\xi < h\} = P\{\xi(y_j) < j \text{ for all } y_j \in \mathbf{Y}(h) \text{ and } \xi(c) < h(c)\}.$$

Similarly, for a continuous and increasing lower boundary

$$P\{\xi > g\} = P\{\xi(x_j) > j \text{ for all } x_j \in \mathbf{X}(g)\}.$$

◇ **Exercise.** Before reading the following proof, it may be convenient to see the simple fact the equality (15.12) is based upon. For this draw a continuous increasing h, choose some two points y_j and y_{j+1}, and verify that if $\xi(y_j) < j$ and $\xi(y_{j+1}) < j+1$, then ξ can not touch h anywhere in between these two points. △

To show that (15.12) is true first note that the left-hand side there involves more random variables than the right-hand side. Therefore, in (15.12) at least the $\leq$ is true. Now show that $\geq$ is also true. Let y_j and y_k, $k \geq j+1$, be two adjacent points in $\mathbf{Y}(h)$, that is,

$$y_j = \ldots = y_{k-1} < y_k.$$

Then $k - 1 < h(t) \leq k$ for $y_j < t \leq y_k$ and $\xi_n(t) \leq \xi(y_k) \leq k-1 < h(t)$ for $y_j < t \leq y_k$. Hence intersection of events $\{\xi(y_k) \leq k-1\}$ and $\{\xi(c) < h(c)\}$ implies $\{\xi(t) < h(t)$ for all $t \in [0,c]\}$. Likewise, intersection of events $\{\xi(x_k) \geq k+1\}$ implies $\{\xi_n(t) > g(t)$ for all $t \in [0,c]\}$. Hence, it follows that in (15.12) $\geq$ is also true.

Let now $s \in [0,c]$ and consider probability

$$Q(s,m) = P\{g(t) < \xi_n(t) < h(t) \text{ for all } t \in [0,s], \ \xi_n(s) = m\}$$

$$\text{for } g(s) < m < h(s).$$

This is the probability that ξ will not touch the boundaries g and h before s and at s will end up in point m.

Denote by $\pi(\cdot, \lambda)$ the Poisson distribution with intensity λ and let $\Delta t_j = t_{j+1} - t_j$. Also let $t_0 = 0$ and $0 < t_1 < \ldots < t_N = c$ be the ordered set of all distinct points from $\mathbf{X}(g) \cup \mathbf{Y}(h) \cup \{c\}$.

The following recursive formulae allow us to calculate the probabilities $Q(s,m)$ at successive points t_j:

for any m, such that $g(t_{j+1}) < m < h(t_{j+1})$,

$$Q(t_{j+1},m) = \sum_{g(t_j) < l \leq m} Q(t_j,l)\pi(m-l,\Delta t_j); \tag{15.13}$$

for the upper boundary only

$$Q(t_{j+1},m) = \sum_{0 \leq l \leq m} Q(t_j,l)\pi(m-l,\Delta t_j).$$

To explain our interest in $Q(s,m)$ note that

$$P\{g < \xi_n < h\} = \sum_{g(1) < m < h(1)} Q(c,m).$$

Also note that, as we know, the distribution of the binomial process, based on L independent random variables with the uniform distribution on $[0, c]$, is equal to the conditional distribution of the standard Poisson process $\xi(t), t \in [0, c]$, given $\xi(c) = L$, and therefore

$$P\{g < z_L < h\} = \frac{Q(c, L)}{\pi(L, c)}.$$

Hence, the recursive formulae (15.13) lead to a recursive, and quick, method to calculate our non-crossing probabilities.

◇ **Exercise.** To see that (15.13) is true remember that increments $\Delta\xi(t_j), j = 0, 1, \ldots, N$, are independent and consider the question: given the process ξ did not cross the boundaries before t_j and reach the level l, how could it evolve from t_j to t_{j+1} to reach the level m? And what is the probability of this? Then use the total probability formula.

You may like to note that the recursion (15.13) is just the Chapman–Kolmogorov forward equation (see, e.g., Feller [1965], ch. X). △

Remark. Let ξ_n be now *any* point process and denote $A_n(t) = \langle \xi_n \rangle(t)$ its compensator with respect to some given filtration $\mathscr{F}_n = \{\mathscr{F}_{tn}, 0 \le t \le 1\}$. One can think of A_n as an "intrinsic time" for ξ_n. For instance, for a Poisson process with intensity n the compensator $A_n(t) \equiv nt$. Hence it is quite natural to consider non-crossing probabilities for ξ_n of the form

$$P\{g_n \circ a_n < \xi_n < h_n \circ a_n\} = P\{g_n(a_n(t)) < \xi_n(t) < h_n(a_n(t)), 0 \le t \le 1\} \tag{15.14}$$

with $a_n(t) = A_n(t)/n$.

But it is well known (see, e.g., Bremaud [1981], Karr [1991], Liptser and Shiryaev [2001]) that with the random change of time the process $\xi_n \circ a_n^{-1}$ is a Poisson process with intensity n, and therefore (15.14) reduces to

$$P\{g_n(s) < \xi_n(a_n^{-1}(s)) < h_n(s), 0 < s < a_n(1)\}$$

It is the typical case, however, that for A_n the (strong) LLN is true:

$a_n(t) \to a(t)$, $0 < t < 1$, as $n \to \infty$ for some deterministic function a. It seems likely then, that the probability in (15.14) can be approximated by the probability

$$P\{g_n(s) < \xi_n \circ a_n^{-1}(s) < h_n(s), \ 0 < s < a(1)\}$$

for which (15.12) works. The proof, however, is needed.

Lecture 16

Population dynamics

Let us first introduce the main components of our story and agree on some notation.

The sequence

$$\{\beta_i\}_{i=1}^{\infty}$$

denotes the times of birth for individuals with given proper-ties/characteristics: for example, individuals with the same place of birth, belonging to a given socio-economic group, of the same eth-nicity, etc. For us, the sequence of times of birth is just an increasing sequence of random variables

$$0 < \beta_1 < \beta_2 < \cdots .$$

Choosing the initial moment of 0 is a matter of agreement: it could be the moment of Adam's creation; or it could be the year 1900, for which β_1 is the moment of the first birth on or after the 1st of January 1900; or if 0 is the 1st of January 2000, then β_1 is the moment of the first birth after this date, and so on.

Consider the point process

$$B(t) = \sum_{i=1}^{\infty} I_{\{\beta_i < t\}},$$

which we will call a birth process and which at any given moment t equals the number of births before time t.

It is worth noting that random moments $\beta_1, \beta_2, \ldots$ are, of course, dependent random variables; compare them with the order statistics in Lecture 3 (in particular, see (3.4)). It is tempting to talk about the "moment of birth of a randomly selected individual" and then postulate the independence of these moments as random variables. However, this could look clumsy: each of these "randomly selected individuals" has to be born at a certain moment β_i, the i-th moment in a sequence, i.e. between moments β_{i-1} and β_{i+1}. Thus, it is essential to start with increasing, and hence dependent, random variables.

Now let γ_i be the moment of death of the individual born at the moment β_i. It is clear that $\gamma_i \geq \beta_i$, and *a priori* we cannot state anything else about the random variables $\gamma_1, \gamma_2, \ldots$. The point process

$$D(t) = \sum_{i=1}^{\infty} I_{\{\gamma_i < t\}} = \sum_{i=1}^{\infty} I_{\{\gamma_i < t\}} I_{\{\beta_i < t\}},$$

called a death process, gives us for any t the number of deceased before the moment t. Then $T_i = \gamma_i - \beta_i$ is the duration of life of the i-th individual, and the assumption of independence of $T_1, T_2, \ldots$ is now quite acceptable (see Lecture 5).

Let us introduce one more process

$$C(t; t_0, \Delta) = C(t) = \sum I_{\{\gamma_i < t + t_0\}} I_{\{t_0 \leq \beta_i < t_0 + \Delta\}},$$

which is the death process for the generation, or cohort, of people born between t_0 and $t_0 + \Delta$. For small Δ, and even Δ equal to 1 year can be regarded as "small" – after all it is just $1/80$ or $1/100$ of the overall "typical" range of values for T_i – we can see that the process $C(t)$ "almost equals" the process

$$C_\Delta(t) = \sum_{j=1}^{N} I_{\{T_j < t\}}, \qquad (16.1)$$

where the cohort size N is

$$N = B(t_0 + \Delta) - B(t_0).$$

We intend to substitute the process $C_\Delta(t)$ for the process $C(t)$.

Often we can reasonably assume that durations of life T_j, $j = 1, \ldots, N$, for members of a given cohort, are not only independent, but also identically distributed. Then, for each given N, the process (16.1) is no different from the binomial process, introduced in Lecture 3, although this fact may be obscured by the different notation.

The process

$$P(t) = B(t) - D(t) \qquad (16.2)$$

defines the population: at the moment t the value of $P(t)$ is the number of individuals out of all those born before the moment t who are still alive. As a difference between two point processes, in queuing theory $P(t)$ is called a queue or storage (see, e.g., Bremaud [1981]).

We can quite easily obtain simple and practically useful models of a population's evolution over time. Indeed, let

$$\mathscr{F}_t = \sigma\{B(s), D(s), s \le t\} \qquad (16.3)$$

be the σ-algebra generated by the processes B and D up till the current moment t (cf. (9.10)), and suppose the distribution of the number of people born during the small interval $[t, t + \Delta)$ and the number of deceased during the same interval under the condition $\mathscr{F}_t$ depends only on the size of the population at moment t and on nothing else. Moreover, suppose

$$
\begin{aligned}
\mathsf{E}[B(t + \Delta) - B(t) \mid \mathscr{F}_t] &= bP(t)\Delta + o(\Delta), \\
\mathsf{E}[D(t + \Delta) - D(t) \mid \mathscr{F}_t] &= dP(t)\Delta + o(\Delta),
\end{aligned}
\qquad (16.4)
$$

so that

$$\mathsf{E}[P(t + \Delta) - P(t) \mid \mathscr{F}_t] = cP(t)\Delta + o(\Delta), \qquad c = b - d.$$

Heuristically speaking, per individual in a given population, b individuals are born and d individuals die over a unit time interval, so that per individual and per unit time, the population is growing on average by c individuals.

Coefficients b and d are called birth and death rates (or birth and death coefficients), while c is called the Malthusian growth rate (the Malthusian coefficient).

Death and birth rates are, clearly, a generic characteristic of the population. The birth rate, in particular, very much depends on a person's age, and, as we know well by now, so does the death rate. The rates we have introduced do not reflect this dependence and rather characterize an abstract "average" person. However, this generality makes them convenient: by relatively simple tools they allow us to reflect the main features in the evolution of a population.

According to Statistics New Zealand, the birth rate in New Zealand in 2007 was 13.61 while the death rate was 7.54, both per thousand of population.

Together with the conditions

$$EB(t) < \infty, \quad ED(t) < \infty, \qquad 0 < t < \infty,$$

equality (16.4) is equivalent to the statement that the random process

$$dM(t) = dP(t) - cP(t)\,dt, \tag{16.5}$$

or, in integral form,

$$M(t) = P(t) - c\int_0^t P(s)\,ds, \tag{16.6}$$

is a martingale with respect to the filtration (16.3):

$$E\left[M(t)\,|\,\mathscr{F}_s\right] = M(s), \quad t \geq s,$$
$$E|M(t)| < \infty.$$

Moreover, according to (16.4), each of the processes

$$M_B(t) = B(t) - b\int_0^t P(s)\,ds,$$
$$M_D(t) = D(t) - d\int_0^t P(s)\,ds$$

is a martingale with respect to the filtration (16.3), and (16.6) is simply their difference.

As we have seen in Lecture 10, the quadratic variation processes of these martingales are equal to the compensators of the point processes B and D, respectively:

$$\langle M_B\rangle(t) = b\int_0^t P(s)\,ds,$$
$$\langle M_D\rangle(t) = d\int_0^t P(s)\,ds.$$

Let us find the quadratic variation of the martingale M, that is, find the process $\langle M \rangle(t)$, which compensates $M^2(t)$ to martingale. We have

$$M^2(t) = M_B^2(t) + M_D^2(t) - 2M_B(t)M_D(t),$$

so that we would have

$$\langle M \rangle(t) = (b+d) \int_0^t P(s)\,ds, \qquad (16.7)$$

if only $M_B(t)M_D(t)$ happened to be a martingale. However, *if the random variables $\{\gamma_i\}_1^\infty$ are continuously distributed, i.e., if*

$$P\{\beta_i = \gamma_j\} = 0 \quad \text{for all } i \text{ and } j, \qquad (16.8)$$

then the process $M_B(t)M_D(t)$ is a martingale with respect to the filtration (16.3).

Indeed, using integration by parts (see, e.g., Karr [1991], ch.2, p.58)

$$M_B(t)M_D(t) = \int_0^t M_B(s-)\,M_D(ds)$$
$$+ \int_0^t M_D(s-)\,M_B(ds) + \sum_{0 \le s \le t} \delta M_B(s)\delta M_D(s), \quad (16.9)$$

where $\delta M_B(s) = M_B(s+) - M_B(s-)$ and $\delta M_D(s) = M_D(s+) - M_D(s-)$ denote the jumps of M_B and M_D at point s, respectively. However, since $\beta_i \ne \gamma_j$ for all i, j with probability 1, the processes M_B and M_D do not have jumps at the same point, so that $\delta M_B(s)\delta M_D(s) = 0$ for all s.

At the same time, both integrals in the right-hand side of (16.9) are martingales. Therefore, the product

$$M_B(t)M_D(t)$$

is indeed a martingale.

One can elaborate on this point a little. Namely, the mutual quadratic variation of martingales $M_1(t)$ and $M_2(t)$, with respect to a given flow of σ-algebras, is a process $\langle M_1, M_2 \rangle(t)$, such that the difference

$$M_1(t)M_2(t) - \langle M_1, M_2 \rangle(t)$$

is again a martingale. If one were to think about $\langle M \rangle(t)$ as a sum of conditional variances, the process $\langle M_1, M_2 \rangle(t)$ could be thought of as the sum of the conditional covariances $\mathsf{E}[dM_1(t)dM_2(t) \mid \mathscr{F}_t]$. Martingales $M_1(t)$ and $M_2(t)$ are called orthogonal if $\langle M_1, M_2 \rangle(t) = 0$ for all t. From the condition (16.8) on their jump points, then, we have shown that the martingales M_B and M_D are orthogonal.

Now let us return to the equality (16.5). Being a definition of M, this equality is simply an identity. However, let us assume that M is given and consider (16.5) as an equation with respect to P: the solution of

$$dP(t) = cP(t)dt + dM(t)$$

is given by

$$P(t) = \int_0^t e^{c(t-s)} dM(s) + e^{ct} P(0). \tag{16.10}$$

This representation describes properties of the population size $P(t)$ as a random process quite well. Indeed, first of all we see that the process

$$P(t)e^{-ct} = \int_0^t e^{-cs} dM(s) + P(0) \tag{16.11}$$

is a martingale, i.e. $P(t)e^{-ct}$ fluctuates around its initial value $P(0)$. But we also see that these fluctuations are not becoming larger with increasing t. More precisely,

there is a random variable Z, with $\mathsf{E}|Z|^2 < \infty$, such that

$$P(t)e^{-ct} \to Z + P(0) \tag{16.12}$$

with probability 1.

To show this, recall that if m_t is a martingale such that

$$\sup_{t \geq 0} \mathsf{E}|m_t| < \infty,$$

then $m_t \to m_\infty$ with probability 1, where $\mathsf{E}|m_\infty| < \infty$ (see, e.g., Liptser and Shiryaev [2001], vol. 1). Therefore to prove (16.12), it is sufficient to show that the martingale $m_t = \int_0^t e^{-cs} dM(s)$, $t \geq 0$, satisfies the aforementioned condition. However, according to the Schwarz inequality

$$\mathsf{E}\left| \int_0^t e^{-cs} dM(s) \right| \leq \left(\mathsf{E}\left[\int_0^t e^{-cs} dM(s) \right]^2 \right)^{1/2}.$$

For the expected value in the right-hand side, using (16.7) we can write

$$\mathsf{E}\left[\int_0^t e^{-cs}\,dM(s)\right]^2 = \mathsf{E}\int_0^t e^{-2cs}\langle M\rangle(ds) = (b+d)\,\mathsf{E}\int_0^t e^{-2cs}P(s)\,ds$$

$$= (b+d)\int_0^t e^{-cs}P(0)\,ds \le \frac{b+d}{c}P(0),$$

so that

$$\sup_{t\ge 0}\mathsf{E}|m_t| < \left(\frac{b+d}{c}\right)^{1/2}P^{1/2}(0)$$

and the sufficient condition for convergence in (16.12) follows.

◇ **Exercise.** Although the fluctuations of $P(t)e^{-ct}$ do not grow with increasing t, it is interesting to look at the behavior of the fluctuation of $P(t)$ itself. According to (16.10) the expected value of $P(t)$ is

$$\overline{P}(t) = \mathsf{E}P(t) = P(0)e^{ct}.$$

Find the variance. How does the variance behave as a function of t? △

The sequences $\{\beta_i\}_{i=1}^\infty$ and $\{\gamma_i\}_{i=1}^\infty$, and hence the processes B, D and P, could be given indices, highlighting additional features of a population. For example, $P_l(t)$, $l = 1,\ldots,m$, may denote a population in the l-th region or country among m different regions, or in the l-th social or ethnic group out of m such groups, and so on. If these populations do not interact and remain closed, nothing new arises: for each $l = 1,\ldots,m$, $P_l = B_l - D_l$, and we simply have m identical unconnected problems. When, however, there are flows between the populations, then (16.2) cannot be used and we have to consider migration.

Assume now that with the i-th individual from the l-th population we associate a random emigration moment κ_{il}, along with a mark z_{il}, which shows the population to which the individual has emigrated. Clearly,

$$\kappa_{il} > \beta_{il}, \qquad i = 1, 2, \ldots, \qquad l = 1, 2, \ldots, m,$$

with probability 1, but we also need to guarantee that $\kappa_{il} \le \gamma_{il}$, or else we could interpret γ_{il} as an emigration moment and introduce one more value of a mark, say, $z_{il} = 0$, which could indicate a migration into an absorbing population, labelled 0. We will, however, preserve different notations κ_{il} and γ_{il}, because mortality rates and migration rates are viewed and studied separately in demography.

Therefore, the point process

$$Q_{lk}(t) = \sum_{i=1}^{\infty} I_{\{\beta_{il}<t\}} I_{\{\gamma_{il}>\kappa_{il}\}} I_{\{\kappa_{il}<t\}} I_{\{z_{il}=k\}}, \quad l \neq k,$$

represents an emigration process from the l-th population into the k-th population; or more briefly, from l to k. For each t the value of $Q_{lk}(t)$ equals the number of those who have emigrated from l to k before time t. The process

$$Q_l(t) = \sum_{k=1}^{m} Q_{lk}(t) = \sum_{i=1}^{\infty} I_{\{\beta_{il}<t\}} I_{\{\kappa_{il}<t\}} I_{\{\gamma_{il}>\kappa_{il}\}}$$

is the process of emigration from l. In contrast to (16.2), we now define the population P_l as

$$P_l(t) = B_l(t) - D_l(t) - \sum_{\substack{k=1 \\ k \neq l}}^{m} Q_{lk}(t) + \sum_{\substack{k=1 \\ k \neq l}}^{m} Q_{kl}(t), \quad l = 1,2,\ldots,m,$$

$$(16.13)$$

and denote by $\mathscr{F}_t^m$ the σ-algebra generated by the birth, death and migration processes in all populations, considered up to the moment t:

$$\mathscr{F}_t^m = \sigma\{B_l(s), D_l(s), Q_{lk}(s), l, k = 1,2,\ldots,m, s \leq t\}.$$

Concerning the distribution of marks, assume that $\{z_{il}\}_{i=1,l=1}^{\infty, m}$ are independent and that the distribution of z_{il} depends only on l, that is, on the population from which the emigration is taking place:

$$P\{z_{il} = k\} = p_l(k), \quad p_l(l) = 0.$$

Assume, moreover, that

$$E\left[B_l(t+\Delta) - B_l(t) \mid \mathscr{F}_t^m\right] = b_l P_l(t)\Delta + o(\Delta),$$

$$E\left[D_l(t+\Delta) - D_l(t) \mid \mathscr{F}_t^m\right] = d_l P_l(t)\Delta + o(\Delta), \qquad (16.14)$$

$$E\left[Q_{lk}(t+\Delta) - Q_{lk}(t) \mid \mathscr{F}_t^m\right] = q_l p_l(k) P_l(t)\Delta + o(\Delta).$$

The intuitive meanings of the coefficients b_l and d_l are the same as previously: they are birth and death rates in the l-th population. The emigration coefficient, or the emigration rate q_l, is equal to the expected number of emigrants per individual in the l-th population during a unit time interval, so that $q_l p_l(k)$ is equal to the expected number of emigrants from l into k per individual in the l-th population per unit time.

We define processes M_l by the equations

$$dP_l(t) = c_l P_l(t)\,ds + \sum_{k=1}^{m} q_k p_k(l) P_k(t)\,dt + dM_l(t), \quad l = 1,2,\dots,m,$$

$$(16.15)$$

where

$$c_l = b_l - d_l - q_l$$

are the effective Malthusian coefficients.

Together with the conditions

$$\mathsf{E}B_l(t) < \infty, \quad \mathsf{E}D_l(t) < \infty, \quad \mathsf{E}Q_{lk}(t) < \infty,$$
$$t < \infty, \quad l,k = 1,2,\dots,m,$$

which mean that none of the point processes under consideration "explodes" prior to any finite moment t, the equalities (16.14) imply that the processes M_l are all martingales.

All of the sequences $\{\beta_{il}\}_{i=1}^{\infty}$, $\{\gamma_{il}\}_{i=1}^{\infty}$ and $\{\kappa_{il}\}_{i=1}^{\infty}$ are jump moments of M_l. In addition, some of the moments $\{\kappa_{ik}\}_{i=1}^{\infty}$ with $k \neq l$, namely those with the marks $z_{ik} = l$, are also jump moments of M_l. Therefore, any two martingales, M_l, M_k, $l \neq k$, have common jump points: κ_{il} with $z_{il} = k$ and κ_{ik} with $z_{ik} = l$, $i = 1,2,\dots$. As a result, the martingales M_l and M_k are not orthogonal. Let us find the matrix of their mutual quadratic variation processes $\langle M_l, M_k \rangle$, $l,k = 1,2,\dots,m$.

Let us start with the diagonal elements $\langle M_l, M_l \rangle = \langle M_l \rangle$. It seems to us natural and not restrictive to assume that the distributions of random variables β_{il}, γ_{il}, κ_{il} for all $i = 1,2\dots$ and $l = 1,2,\dots,m$, are continuous, so that

$$\mathsf{P}\{\beta_{il} = \gamma_{jk}\} = \mathsf{P}\{\beta_{il} = \kappa_{jk}\} = \mathsf{P}\{\gamma_{ik} = \kappa_{jk}\} = 0,$$
$$\mathsf{P}\{\kappa_{il} = \kappa_{jk}\} = 0, \quad (i,l) \neq (j,k),$$
$$i,j = 1,2,\dots, \quad k,l = 1,2,\dots,m.$$
$$(16.16)$$

It is true that if we start rounding exact dates to months or years, then there will be many coinciding moments; but this is a matter of data manipulation, and is not intrinsic to the phenomenon being modelled. For $\langle M_l \rangle$ we obtain:

if the assumption (16.16) is satisfied then

$$\langle M_l \rangle(t) = (b_l + d_l + q_l) \int_0^t P_l(s)\,ds + \sum_{k=1}^{m} q_k p_k(l) \int_0^t P_k(s)\,ds, \quad (16.17)$$

where, we recall, $p_l(l) = 0$.

The statement quickly follows from the following lemma. This lemma explains the form not only of expression (16.17) but also of the whole matrix $\langle M_l, M_k \rangle$ below. The lemma states:

if assumption (16.16) is satisfied, then the martingales

$$M_{Bl}(t) = B_l(t) - b_l \int_0^t P_l(s)\,ds,$$

$$M_{Dl}(t) = D_l(t) - d_l \int_0^t P_l(s)\,ds, \qquad l,k = 1,\ldots,m,$$

$$M_{Qlk}(t) = Q_{lk}(t) - q_l p_l(k) \int_0^t P_l(s)\,ds,$$

are orthogonal; their quadratic variations are given by

$$\langle M_{Bl} \rangle(t) = b_l \int_0^t P_l(s)\,ds,$$

$$\langle M_{Dl} \rangle(t) = d_l \int_0^t P_l(s)\,ds,$$

$$\langle M_{Qlk} \rangle(t) = q_l p_l(k) \int_0^t P_l(s)\,ds.$$

This lemma is a direct extension of the result on martingales M_B and M_D above; see the discussion following (16.8). On the other hand, it is clear that

$$M_l(t) = M_{Bl}(t) - M_{Dl}(t) - \sum_{\substack{k=1\\k\neq l}}^{m} M_{Qlk}(t) + \sum_{\substack{k=1\\k\neq l}}^{m} M_{Qkl}(t) \qquad (16.18)$$

and because the summands are orthogonal, the quadratic variation

$\langle M_l \rangle(t)$ is simply a sum of the quadratic variation of the summands:

$$\langle M_l \rangle(t) = \langle M_{Bl} \rangle(t) + \langle M_{Dl} \rangle(t) + \sum_{\substack{k=1 \\ k \neq l}}^{m} \langle M_{Qlk} \rangle(t) + \sum_{\substack{k=1 \\ k \neq l}}^{m} \langle M_{Qkl} \rangle(t),$$

which is (16.17).

For $\langle M_l, M_k \rangle(t)$ we have:

if assumption (16.16) *is satisfied, then for* $l \neq k$

$$\langle M_l, M_k \rangle(t) = q_l p_l(k) \int_0^t P_l(s)\,ds + q_k p_k(l) \int_0^t P_k(s)\,ds. \qquad (16.19)$$

Indeed, according to (16.18) the product of martingales $M_l(t)M_k(t)$ equals a sum of products of orthogonal martingales and two squares, $M_{Qlk}^2(t)$ and $M_{Qkl}^2(t)$. Therefore

$$\langle M_l, M_k \rangle(t) = \langle M_{Qlk} \rangle(t) + \langle M_{Qkl} \rangle(t),$$

which is (16.19).

◇ **Exercise.** Prove the lemma on orthogonality of the martingales M_{Bl}, M_{Dl}, M_{Qlk}, $l, k = 1, \ldots, m$. One will only need to repeat the argument for orthogonality of M_B and M_D from the first part of this lecture. △

So, now we agree that populations P_l, $l = 1, \ldots, m$, defined according to equations (16.13), satisfy the system of linear stochastic differential equations (16.15), which follow from assumptions (16.14). Moreover, we know the structure of its "noisy" martingale components M_l, $l = 1, \ldots, m$.

Using the matrix $\mathbf{C}$:

$$\mathbf{C} = (c_{lk}), \quad c_{ll} = c_l, \quad c_{kl} = q_k p_k(l),$$

and vector notations $\mathbf{P}(t)$ and $\mathbf{M}(t)$

$$\mathbf{P}(t) = \begin{pmatrix} P_1(t) \\ \vdots \\ P_l(t) \end{pmatrix}, \qquad \mathbf{M}(t) = \begin{pmatrix} M_1(t) \\ \vdots \\ M_l(t) \end{pmatrix},$$

the system of equations (16.15) can be re-written as

$$d\mathbf{P}(t) = \mathbf{C}\mathbf{P}(t)dt + d\mathbf{M}(t),$$

or as

$$\mathbf{P}(t) = \int_0^t e^{(t-s)\mathbf{C}}\mathbf{M}(ds) + e^{t\mathbf{C}}\mathbf{P}(0), \qquad (16.20)$$

which is not only similar in form to (16.10), but also to (14.11) of Lecture 14. In Lecture 14 we were considering deterministic (non-random) vector functions P. However, in this lecture too one could be interested in expected populations $\bar{P}_l(t) = \mathsf{E}P_l(t)$. Since $\mathsf{E}M_l(t) = 0$, these expected populations satisfy the system of equations

$$d\bar{P}_l(t) = c_l \bar{P}_l(t)\,dt + \sum q_k p_k(l)\bar{P}_k(t)dt, \qquad l = 1,\ldots,m,$$

or

$$d\bar{\mathbf{P}}(t) = \mathbf{C}\bar{\mathbf{P}}(t)\,dt,$$

of which the solution is

$$\bar{\mathbf{P}}(t) = e^{t\mathbf{C}}\bar{\mathbf{P}}(0).$$

This is a well-known model for mean, or expected, populations. In this context the matrix $\mathbf{C}$ is called a migration matrix. Although the heuristic context of this lecture and of Lecture 14 is very different, the mathematical framework is the same in each case. By the way, in our current lecture too it would be natural to introduce a free term $\mathbf{a}(t)$: if there were an inflow into the populations $P_1, P_2, \ldots, P_m$ from outside, and therefore the system of our populations were not closed, then the following system of equations would arise:

$$d\bar{\mathbf{P}}(t) = \mathbf{C}\bar{\mathbf{P}}(t)\,dt + \mathbf{a}(t)\,dt$$

with the solution

$$\bar{\mathbf{P}}(t) = \int_0^t e^{(t-s)\mathbf{C}}\mathbf{a}(s)\,ds + e^{t\mathbf{C}}\bar{\mathbf{P}}(0), \qquad (16.21)$$

which coincides with (14.9) and (14.10).

Coordinates $a_l(t)\,dt$ of the vector-function $\mathbf{a}(t)dt$ equal the number of immigrants from outside populations $P_1, \ldots, P_m$ into the l-th population over a time interval dt. Apart from qualitative descriptions of the behavior of the populations, could the expressions (16.21) be used for calculation of $\overline{P}(t)$? It seems to us that in many cases they could. For example, for one population (one country)

$$\overline{P}(t) = \int_0^t e^{(t-s)c} a(s) \, ds + e^{tc} \overline{P}(0), \qquad (16.22)$$

where $a(s)$ is the rate of legal immigration into the country. Government organizations use rigorous measures to evaluate and keep track of $a(s)$. Comparison of population sizes, derived according to (16.22), with a real population can give a fair understanding of the size of illegal immigration.

◇ **Exercise.** What will happen in (16.20), if we describe immigration into populations $1, \ldots, m$ from outside as a vector random process with coordinates

$$dQ_l(t) = a_l(t) \, dt + dX_l(t),$$

where X_l, $l = 1, 2, \ldots, m$, are martingales? △

In conclusion, let us make some remarks of a historical and methodological nature.

The classical framework for the description and analysis of population dynamics has always been the domain of the so-called branching processes. The theory originated in Galton and Watson [1874]. One can find a short but very interesting historical review of this theory in the Historical Prologue in Mode [1971], ch. 1.

In a simple model of just one type of individuals, suppose that the i-th member, out of N_n members of the n-th generation, will produce X_{in} offspring, $X_{in} \geq 0, i = 1, 2, \ldots, N_n$, so that the $(n+1)$-st generation will comprise

$$N_{n+1} = \sum_{i=1}^{N_n} X_{in} \qquad (16.23)$$

members. Suppose $N_0 > 0$ and assume all random variables X_{in}, $i \geq 1$,

$n \geq 0$, independent and identically distributed. Set $m = EX_{in}$ and denote by $\mathscr{F}_n$ the σ-algebra generated by all of the past:

$$\mathscr{F}_n = \sigma\{X_{ik}, \ i = 1, \ldots, N_k, \ k = 0, 1, \ldots, n-1\}.$$

Recall that if random variables $\eta_n, n = 1, 2, \ldots$, form a martingale with respect to some filtration, then $\xi_n = \eta_{n+1} - \eta_n, n = 1, 2, \ldots$, form a martingale difference with respect to the same filtration, so we can write

$$N_{n+1} = mN_n + \xi_n, \qquad \xi_n = \sum_{i=1}^{N_n}(X_{in} - m),$$

or

$$N_{n+1} - N_n = (m-1)N_n + \xi_n, \qquad \xi_n = \sum_{i=1}^{N_n}(X_{in} - m), \qquad (16.24)$$

where the ξ_n, $n = 0, 1, 2, \ldots$, form a martingale difference:

$$E[\xi_n \,|\, \mathscr{F}_{n-1}] = 0.$$

The representation (16.24) is analogous to the equality (16.5) for the population $P(t)$, considered earlier; and the coefficient $m-1$ is, essentially, the Malthusian coefficient. Either by solving the difference equation (16.24), or by verifying the answer by direct substitution, which is easier, we can find that the sequence

$$N_n m^{-n} = \sum_{k=1}^{n} \xi_k m^{-k}, \qquad n = 1, 2, \ldots,$$

is a non-negative martingale; and this, again, is analogous to the expression (16.11) for $P(t)e^{-ct}$.

The sequence N_n defined by (16.23) is the well-known Galton–Watson process, of which there are many generalizations for the case of not one but many different types of individuals/particles, for which we refer again to Mode [1971].

Clearly, the Galton–Watson process, as defined in (16.23), describes the dynamics of a size of a generation, and not a population as such. Representatives of the same generation live for different lengths of time and a population consists of representatives from different generations. The extension of the Galton–Watson process to the case in which individuals/particles live for random times is called the Bellman–Harris process: see, in particular, the monograph Harris [1963]. As it was initially introduced in Bellman and Harris [1948], it was assumed that an individual person or particle has an exponentially distributed lifetime and that the moment of birth of a child coincides with the death of the parent. Although suitable for the physical problems of disintegration, these conditions were somewhat restrictive for biological populations. Already in 1971, however, in the monograph Mode [1971] cited above, the theory of branching processes was presented in a completely general form.

Finally, we refer to the monograph Asmussen and Hering [1983], especially in connection with the so-called critical case $m < 1$; and to the monograph Liemant et al. [1988], which studies very interesting general problems of the theory, which however fall beyond the scope of this lecture.

16.1* Age structure of a population

In the same notations as before, consider the birth moment β_i of the i-th born individual along with the corresponding lifetime T_i. Then, at time t the age of this individual will be $t - \beta_i$ and

$$I_{\{\beta_i < t, T_i > t - \beta_i\}}$$

is, certainly, an indicator function of the event that the i-th individual was born before t and is still alive; we, however, will be interested in the indicator function

$$I_{\{\beta_i < t - x, T_i > t - \beta_i\}}$$

of the more general event that the age of an individual alive at t is greater than x. Consider the sum

$$P(t,x) = \sum_{i=1}^{\infty} I_{\{\beta_i < t - x, T_i > t - \beta_i\}}. \tag{16.25}$$

This is an object of interest for us now. It is equal to the number of individuals in the population, who are of age greater than x. For $x = 0$

it is just the population we considered before. The ratio

$$\frac{P(t,x)}{P(t)}$$

describes the age structure of the population. We do not study here probabilistic properties of $P(t,x)$, but will only consider the behavior of its expected value $\mathsf{E}P(t,x)$ as a function of x and the behavior of the ratio

$$\frac{\mathsf{E}P(t,x)}{\mathsf{E}P(t)},$$

and connect it with the lifetime distributions $G_\beta(x)$ and other demographic characteristics.

The expression (16.25) does not incorporate the migration in and out of the population, so that we consider a closed population. But if we do not, the presentation will become rather heavy. It may be not very useful to strive for complete generality all the time.

First consider the conditional expected value of $P(t,x)$ given all birth moments before t. That is, if again $B(t), t \geq 0$, is the birth process, and if

$$\mathscr{F}_t^B = \sigma\{B(s), s \leq t\}$$

is a σ-algebra generated by B up to time t, then we want to consider $\mathsf{E}[P(t,x)|\mathscr{F}_t^B]$. This involves expectation with respect to the T_i-s only. Suppose that the distribution of T_i does not depend on any other pairs $\{\beta_j, T_j\}, j \neq i$, but may depend on the birth time β_i. Since the notation F, with and without indices has been used so far a little bit too often, we will use the notation $G_\beta(x)$ for this distribution function. Then

$$\mathsf{E}[P(t,x)|\mathscr{F}_t^B] = \sum_{i=1}^\infty I_{\{\beta_i < t-x\}}[1 - G_{\beta_i}(t - \beta_i)].$$

It is more convenient to write the sum here as the integral

$$\mathsf{E}[P(t,x)|\mathscr{F}_t^B] = \int_0^{t-x} [1 - G_y(t-y)]\,dB(y). \tag{16.26}$$

Now take an expected value from both sides of (16.26) to obtain

$$\bar{P}(t,x) = \mathsf{E}P(t,x) = \int_0^{t-x} [1 - G_y(t-y)]\,d\bar{B}(y), \tag{16.27}$$

where $\bar{B}(y) = EB(y)$.

For $x = 0$ this leads to an expression for the expected population, somewhat different from what we obtained in the lecture. It now incorporates the age as a part of the model.

The ratio $\bar{P}(t,x)/\bar{P}(t)$ has all the properties of a tail of a distribution function, or $1 - \bar{P}(t,x)/\bar{P}(t)$ has all the properties of a distribution function: it is 0 at $x = 0$, it is non-decreasing in x and it converges to 1 when $x \to \infty$. Instead of the fraction $\bar{P}(t,x)/\bar{P}(t)$ let us consider a more traditional and visually more informative quantity, which is the density of this distribution function:

$$-\frac{1}{\bar{P}(t)}\frac{d_x\bar{P}(t,x)}{dx}.$$

Using the expression in (16.27) one can represent it as

$$-\frac{1}{\bar{P}(t)}\frac{d_x\bar{P}(t,x)}{dx} = \frac{[1 - G_{t-x}(x)]}{\bar{P}(t)}\frac{d_t\bar{B}(t-x)}{dt}$$

$$= [1 - G_{t-x}(x)]\frac{\bar{P}(t-x)}{\bar{P}(t)}b(t-x), \qquad (16.28)$$

where

$$b(t-x) = \frac{1}{\bar{P}(t-x)}\frac{d_x\bar{B}(t-x)}{dt}$$

is the Malthusian birth rate as we introduced it in (16.4). However, now we do not have to assume that it is constant in time.

If we do assume that $b(t-x)$ is constant, it would be a one form to assume that a population is "stationary". Another stationarity assumption would be to say that $G_y(x)$ does not depend on the cohort, that is, on y. What remains is to assume that the expected population stays constant, but we will only assume that $\bar{P}(t)$ is not decreasing. It is a remarkable empirical fact, that in many cases these assumptions cannot be true.

Indeed, if they were, and if we switch back to the notation F for G_y not depending on y, then

$$-\frac{1}{\bar{P}(t)}\frac{d_x\bar{P}(t,x)}{dx} = [1 - F(x)]\frac{\bar{P}(t-x)}{\bar{P}(t)}b, \qquad (16.29)$$

and here $1 - F(x)$ is a decreasing function and the ratio $\bar{P}(t-x)/\bar{P}(t)$ is a not increasing function of x, and therefore, for every t, the ratio

$$-\frac{1}{\bar{P}(t)}\frac{d_x\bar{P}(t,x)}{dx}$$

has to be a decreasing function in x. However, the empirical analog of this ratio, called an age diagram or an age pyramid, in many populations is not a decreasing function.

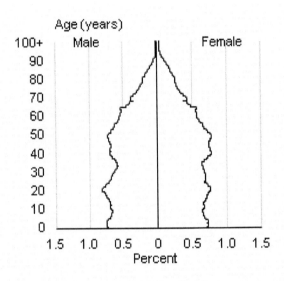

Figure 16.1 *Age pyramid for New Zealand in 2011. Source: Statistics New Zealand.*

In practice, they replace the derivative above by the ratio of increments $\Delta B(t-x) = [B(t-x+\Delta x) - B(t-x)]$ and Δx, with increment of x taken usually as one year, or 5 years, and x being an age in years. In other words, they replace the derivative by the estimated birth rate for the year $t-x$. Also the distribution function F is replaced by its estimation $\tilde{F}$ from mortality tables. The resulting quantity

$$\frac{1-\tilde{F}(x)}{P(t)}\frac{\Delta B(t-x)}{\Delta x},$$

with $\Delta x = 1$, is usually called an "age pyramid", especially when two such functions in x are juxtaposed, with tails of estimated distribution functions $1 - \tilde{F}_m(x)$ for males on one side and $1 - \tilde{F}_f(x)$ for females, in the population, on the other side.

Extremely interesting age pyramids for the population of France in several years from the second half of the 18th century to the mid-20th century are discussed in Pressat [1972].

Natural but simple assumptions we had about the distribution of marks T_i of the marked point process $\{(\beta_i, T_i)\}_{i=1}^{\infty}$ makes expression (16.25) a convenient starting point to study the behavior of

$$P(t,x) - \frac{\bar{P}(t,x)}{\bar{P}(x)} P(t)$$

as a process in x and related problems. This we hope to consider in the part 2 of the total project.

About behavior of the expected population $\bar{P}(t,x)$ in t one can learn much, e.g., from Caswell [2001] and Pollard [1973].

Bibliography

O O Aalen, P K Anderson, O Borgan, R D Gill, and N Keiding. History of application of martingales in survival analysis. *Electronic Journal for History of Probability and Statistics*, 5:28, 2009.

M Abramowitz and I E Stegun. *Handbook of Mathematical Functions*. Dover, 1964.

P K Andersen, O Borgan, R D Gill, and N Keiding. *Statistical Models Based on Counting Processes*. Springer-Verlag, 1993.

S Asmussen and H Hering. *Branching Processes*, volume 3 of *Progress in Probability and Statistics*. Birkhäuser, 1983.

R H Baayen. *Word Frequency Distributions*, volume 18 of *Text, Speech and Language Technology*. Kluwer Academic Publishers, 2002.

R E Barlow and F Proshan. *Statistical Theory of Reliability and Life Testing. Probability Models*. International Series in Decision Processes. Holt, Rinehart and Winston, 1975.

R Bellman and T E Harris. On the theory of age-dependent stochastic branching processes. *Proceedings of the National Academy of Science*, 34:601–604, 1948.

P Billingsley. *Convergence of Probability Measures*. Wiley, 1977.

L N Bol'shev and N V Smirnov. *Tables of mathematical statistics*. Nauka, 1983.

P Bremaud. *Point Processes and Queues: Martingale Dynamics*. Springer Series in Statistics. Springer-Verlag, 1981.

N Breslow and J Crowley. A large sample study of the life table and product limit estimates under random censorship. *Annals of Statistics*, 2:437–453, 1970.

M Bulgakov. *The Master and Margarita*. Alfred A Knopf, 1992. Trans. by Michael Glenny.

H Caswell. *Matrix Population Models*. Sinauer Associates Inc, 2001.

D R Cox. Regression models and life tables (with discussion). *Journal of the Royal Statistical Society, Series B*, 34:187–220, 1972.

H Cramér. *Mathematical Methods of Statistics*. Princeton University Press, 1946.

H Cramér. On the history of certain expansions used in mathematical statistics. *Biometrika*, 59(1):205–207, 1972.

L de Haan and A Ferreira. *Extreme Value Theory: An Introduction*. Springer-Verlag, 2006.

M Denuit, C Lefevre, and M Mesfioui. Stochastic orderings of convex-type for discrete bivariate risks. *Scandinavian Actuarial Journal*, 1999(1):32–51, 1999.

J L Doob. Heuristic approach to the Kolmogorov–Smirnov theorems. *Annals of Mathematical Statistics*, 20:393–403, 1949.

K Dunstan, A Howard, and J Cheung. A History of Survival in New Zealand: Cohort Life Tables 1876–2004. Technical report, Statistics New Zealand, 2006. Wellington.

J Durbin. Weak convergence of the sample distribution function when parameters are estimated. *Annals of Statistics*, 1(2):279–290, 1973.

W Feller. *An Introduction to Probability Theory and its Applications*, volume 1. Wiley, 1965.

W Feller. *An Introduction to Probability Theory and its Applications*, volume 2. Wiley, 1971.

F Galton and H W Watson. On the probability of extinction of families. *Anthropological Institute of Great Britain and Ireland*, 4:138–144, 1874.

E A Gehan. A generalized two-sample Wilcoxon test for doubly censored data. *Biometrika*, 52(3/4):650–653, December 1965.

H U Gerber. *An Introduction to Mathematical Risk Theory*. Huebner Foundation Monograph Series No. 8. Irwin, 1979.

E Gibbon. *The History of the Decline and Fall of the Roman Empire*, volume 1. 7th printing in 1998, Folio Society, 1776.

I I Gikhman. On a problem in the theory of the ω^2 test. *Mathemat. Sbornik of Kiev State University*, 5:51–59, 1954. In Ukrainian.

R D Gill. *Censoring and Stochastic Integrals.* Number 124 in Mathematical Centre Tracts. Mathematisch Centrum, Amsterdam, 1980.

I M Glazman and Ju I Ljubich. *Finite-Dimensional Linear Analysis: A Systematic Presentation in Problem Form.* MIT Press, 1974.

V I Glivenko. Sulla determinazione empirica delle leggi di probabilita. *Giornale dell'Istituto Italiano degli Attuari,* 4:92–99, 1933.

B Gompertz. On the nature of the function expressive of the law of human mortality, and on a new mode of determining the value of life contingencies. *Philosophical Transactions of the Royal Society of London,* 115:513–585, 1825.

M J Goovaerts, F De Vylder, and J Haezendonck. *Insurance Premiums.* North-Holland Publishing Co., Amsterdam, 1984.

P J Green and B W Silverman. *Nonparametric Regression and Generalized Linear Models: A Roughness Penalty Approach.* Chapman and Hall, London, 1994.

L G Gvanceladze and D M Chibisov. On tests of fit based on group data. In *Contributions to Statistics,* pages 79–89. Reidel Dordrecht, 1978.

R P Gwinn, P B Norton, and P W Goetz, editors. *China,* volume 3. Encyclopaedia Britannica Inc., 1989a.

R P Gwinn, P B Norton, and P W Goetz, editors. *Life Span,* volume 7. Encyclopaedia Britannica Inc., 1989b.

S J Haberman and T A Sibbett. *History of Actuarial Science.* Pickering & Chatto, 1995.

W Härdle. *Smoothing Techniques: With Implementation in S.* Springer-Verlag, Berlin, 1991.

T E Harris. *The Theory of Branching Processes.* Springer-Verlag, 1963.

J Haywood and E V Khmaladze. On distribution-free goodness-of-fit testing of exponentiality. *Journal of Econometrics,* 143:5–18, 2008.

J Haywood and E V Khmaladze. Testing exponentiality of distribution. In *International Encyclopedia of Statistical Science,* pages 1587–1590. Springer-Verlag, 2011.

P J Huber. *Robust Statistics.* Wiley, 1981.

G Ishii. On the exact probabilities of Rényi's test. *Annals of the Institute of Statistical Mathematics*, 2:13–34, 1959.

N L Johnson, S Kotz, and N Balakrishnan. *Continuous Univariate Distributions*, volume 1. Wiley, 2nd edition, 1994.

N L Johnson, S Kotz, and N Balakrishnan. *Continuous Univariate Distributions*, volume 2. Wiley, 2nd edition, 1995.

C B Jordan. Census of the maori population. *Journals of the House of Representatives of New Zealand (A to J's)*, Appendix, Session II, H39A:1–8, 1921.

M Kac, J Kiefer, and J Wolfowitz. On test of normality and other tests of goodness of fit based on distance methods. *Annals of Mathematical Statistics*, 26:189–211, 1955.

A F Karr. *Point Processes and their Statistical Inference*, volume 7 of *Probability: Pure and Applied*. Marcel Dekker, 2nd edition, 1991.

E V Khmaladze. The use of ω^2-type tests for testing parametric hypothesis. *Theory of Probability and its Applications*, 24(2):283–301, 1979.

E V Khmaladze. Martingale approach in the theory of goodness-of-fit tests. *Theory of Probability and its Applications*, 26(2):240–257, 1981. Translated by A B Aries.

E V Khmaladze. Comment on 'Sulla determinazione empirica di una legge di distribuzione'. In A N Shiryaev, editor, *Selected Works of A N Kolmogorov Volume II: Probability Theory and Mathematical Statistics*, number 26 in Mathematics and its Applications (Soviet Series), pages 574–582. Kluwer Academic Publishers, 1992.

E V Khmaladze. Goodness of fit problem and scanning innovation martingales. *Annals of Statistics*, 21(2):798–829, 1993.

E V Khmaladze. Zipf's law. In *Encyclopedia of Mathematics*, pages 460–463. Kluwer, 2001.

E V Khmaladze and E Shinjikashvili. Calculation of non-crossing probabilities for poisson processes and its corollaries. *Advances in Applied Probability*, 33(3):702–716, September 2001.

E V Khmaladze, R Brownrigg, and J Haywood. Brittle power: On Roman emperors and exponential lengths of rule. *Statistics & Probability Letters*, 77(12):1248–1257, 2007.

E V Khmaladze, R Brownrigg, and J Haywood. Memoryless reigns of the "Sons of Heaven". *International Statistical Review*, 78(3): 348–362, 2010.

D Kienast. *Römische Kaisertabelle: Grundzüge Römischen Kaiserchronologie*. Wissenschaftliche Buchgesellschaft, Darmstadt, 1990.

F C Klebaner. *Introduction to Stochastic Calculus with Applications*. Imperial College Press, 2005.

R Koenker. *Quantile Regression*, volume 38 of *Econometric Society Monographs*. Cambridge University Press, 2005.

R Koenker and Z Xiao. Unit root quantile autoregression inference. *Journal of the American Statistical Association*, 99(467):775–787, 2004.

A N Kolmogorov. Sulla determinazione empirica di una legge di distribuzione. *Giornale dell'Istituto Italiano degli Attuari*, 4:83–91, 1933.

A N Kolmogorov. *Selected Works of A N Kolmogorov, Volume II: Probability Theory and Mathematical Statistics*. Number 26 in Mathematics and its Applications (Soviet Series). Kluwer Academic Publishers, 1992.

E D Kopf. The early history of annuity. *Proceedings of the Casualty and Actuarial Society*, XIII:225–266, 1926.

H L Koul and L Sakhanenko. Goodness of fit testing in regression: a finite sample comparison of bootstrap methodology and khmaladze transformation. *Statistics & Probability Letters*, 74(3):290–302, 2005.

H L Koul and E Swordson. Khmaladze transformation. In *International Encyclopedia of Statistical Science*, pages 715–718. Springer-Verlag, 2010.

J Landers. *Death and the Metropolis: Studies in the Demographic History of London 1670-1830*. Cambridge University Press, 1993.

A Liemant, A Wakolbinger, and K Matthes. *Equilibrium Distributions of Branching Processes*. Kluwer Academic Publishers, 1988.

R Liptser and A N Shiryaev. *Statistics of Random Processes*. Springer-Verlag, 2001.

M Lovric, editor. *International Encyclopedia of Statistical Science.* Springer-Verlag, 2011. In three volumes.

N Maglapheridze, Z P Tsigroshvili, and M van Pul. Goodness-of-fit tests for parametric hypotheses on the distribution of point processes. *Mathematical Methods of Statistics*, 1989.

E Marsden and T Barratt. The probability distribution of the time intervals of particles with applications to the number of particles emitted by uranium. *Proceedings of Physical Society of London*, 23:377, 1910.

E Marsden and T Barratt. The α-particles emitted by active deposits of thorium and actinium. *Proceedings of Physical Society of London*, 24:50, 1911.

G V Martynov. *Omega-square Goodness of Fit Test.* Nauka, 1978.

Ch J Mode. *Multitype Branching Processes. Theory and Applications.* American Elsevier Publishing Company, 1971.

D S Moore. A chi-square statistic with random cell boundaries. *Annals of Mathematical Statistics*, 42(1):147–156, 1971.

E P Neale. The 1936 maori census. *Economic Record*, 16(2):275–280, 1940.

A Neill. *Life Contingencies.* Heinemann, 1989.

A Nikabadze and W Stute. Model checks under random censorship. *Statistics & Probability Letters*, 32(3):249–259, 1977.

M S Nikulin. The chi-square test for continuous distributions with location and scale parameters. *Theory of Probability and its Applications*, 18(3):559–568, 1973.

J H Pollard. *Mathematical Models for the Growth of Human Population.* Cambridge University Press, 1973.

R Pressat. *Population.* Penguin Books Inc, 1970.

R Pressat. *Demographic Analysis: Methods, Results, Applications.* Aldine - Atherton, 1972.

R Rebolledo. Central limit theorems for local martingales. *Zeitschrift fur Wahrscheinlichkeitstheorie und verwandte Gebiete*, 51:269–286, 1980.

R D Reiss and M Thomas. *Statistical Analysis of Extreme Values with*

Applications to Insurance, Finance, Hydrology and Other Fields. Birkhaüser-Verlag, 3rd edition, 2007.

E Rutherford, J Chadwick, and C D Ellis. *Radiations from Radioactive Substances.* Cambridge University Press, 1930.

A N Shiryaev. *Probability.* Springer-Verlag, 1980.

A N Shiryaev. *Essentials of Stochastic Finance: Facts, Models, Theory.* Advanced Series on Statistical Science and Applied Probability. World Scientific, 1999.

G R Shorack and J A Wellner. *Empirical Processes with Applications to Statistics.* Wiley, 2009.

B W Silverman. *Density Estimation for Statistics and Data Analysis.* Chapman and Hall, London, 1986.

N V Smirnov. Probabilities of large values of non-parametric one-sided goodness of tests. *Proceeding of Mathematical Institute of the USSR Academy of Sciences*, 64:185–210, 1961.

W Stute and J-L Wang. The strong law under random censorship. *Annals of Statistics*, 21(3):1591–1607, 1993.

A W van der Vaart and J A Wellner. *Weak Convergence and Empirical Processes with Applications to Statistics.* Springer-Verlag, 1996.

E J Vaughan and T M Vaughan. *Fundamentals of Risk and Insurance.* Wiley, 2008.

M P Wand and M C Jones. *Kernel Smoothing.* Chapman and Hall, London, 1995.

W Weibull. A statistical representation of fatigue failure in solids. *Kunglig Tekniska Högskolans Hamdlingar*, 27, 1939.

G K Zipf. *Human Behaviour and the Principles of Least Effort.* Addison-Wesley, 1949.

Index

actuarial value of a contract, 152
age pyramid, 214
age structure, 212
annuity
    deferred annuity, 164
    life annuity, 163
    term annuity, 164
    term deferred annuity, 165

Balkema–de Haan–Pickands
    theorem, 187
Berry–Esséen inequality, 34
beta distribution, 20
beta function, 20
binomial distribution, 25
binomial process, 23
birth process, 197
bonus, 149
    compound reversionary, 149, 150
    simple reversionary, 149
Brownian bridge, 46
    as projection of Brownian motion, 45
    in time F, 45
    standard, 45
Brownian motion
    standard, 43
    with respect to time F, 43

censoring observations, 97

central limit theorem
    for martingales, 111
    for multinomial random vector, 38
characteristic function of multinomial distribution, 36
Chebyshev inequality, 27
chi-square distribution function, 16
chi-square goodness of fit statistic, 48
    limit distribution of, 51
Chinese emperors
    duration of rule, 71
cohort, 57
compensator, 102
confidence bound
    for the tail of distribution function, 176
    for expected remaining life, 186
confidence limits for F, 59
convergence in distribution, 47
costs of future payments at initial moment, 143
covariance function
    of Brownian motion in time F, 44

For Product Safety Concerns and Information please contact our EU
representative GPSR@taylorandfrancis.com
Taylor & Francis Verlag GmbH, Kaufingerstraße 24, 80331 München, Germany

www.ingramcontent.com/pod-product-compliance
Ingram Content Group UK Ltd.
Pitfield, Milton Keynes, MK11 3LW, UK
UKHW021615240425
457818UK00018B/566